Lecture Notes in Artificial Intell

Edited by J. G. Carbonell and J. Siekmann

Subseries of Lecture Notes in Computer Science

Thomas R. Roth-Berghofer   Stefan Schulz
David B. Leake (Eds.)

Modeling and Retrieval of Context

Second International Workshop, MRC 2005
Edinburgh, UK, July 31–August 1, 2005
Revised Selected Papers

 Springer

Series Editors

Jaime G. Carbonell, Carnegie Mellon University, Pittsburgh, PA, USA
Jörg Siekmann, University of Saarland, Saarbrücken, Germany

Volume Editors

Thomas R. Roth-Berghofer
Deutsches Forschungszentrum für Künstliche Intelligenz DFKI GmbH
Erwin-Schrödinger-Straße 57, 67663 Kaiserslautern, Germany
E-mail: thomas.roth-berghofer@dfki.de

Stefan Schulz
The e-Spirit Company GmbH
Barcelonaweg 14, 44269 Dortmund, Germany
E-mail: schulz@e-spirit.de

David B. Leake
Indiana University
Computer Science Department
Lindley Hall 215, 150 S. Woodlawn Avenue, Bloomington, IN 47405-7104, USA
E-mail: leake@cs.indiana.edu

Library of Congress Control Number: 2006923358

CR Subject Classification (1998): I.2, F.4.1, J.3, J.4

LNCS Sublibrary: SL 7 – Artificial Intelligence

ISSN   0302-9743
ISBN-10   3-540-33587-0 Springer Berlin Heidelberg New York
ISBN-13   978-3-540-33587-0 Springer Berlin Heidelberg New York

Springer is a part of Springer Science+Business Media

springer.com

© Springer-Verlag Berlin Heidelberg 2006
Printed in Germany

Typesetting: Camera-ready by author, data conversion by Scientific Publishing Services, Chennai, India
Printed on acid-free paper   SPIN: 11740674   06/3142   5 4 3 2 1 0

Preface

Computing in context has become a necessity in modern and intelligent IT applications. With the use of mobile devices and current research on ubiquitous computing, context-awareness has become a major issue. However, context and context-awareness are crucial not only for mobile and ubiquitous computing. They are also vital for spanning various application areas, such as collaborative software and Web engineering, personal digital assistants and peer-to-peer information sharing, health care workflow and patient control, and adaptive games and e-learning solutions. In these areas, context serves as a major source for reasoning, decision making, and adaptation, as it covers not only application knowledge but also environmental knowledge. Likewise, modeling and retrieving context is an important part of modern knowledge management processes.

In addition, context can play a role in determining what information a system should provide. This is important for supporting the users of automated or intelligent systems, for tasks such as explaining how solutions are found, what the system is doing, and why it operates in a certain way. The methods applied and the advice given have to be explained, so that the user can understand the process and agree on decisions. Context is equally important for deciding when to provide uncertain or blurred information, e.g., when using a tracking system in situations for which either revealing the current position, or denying access to it, would have adverse effects.

In this wide range of applications, context is now more than just location. It is seen as a multi-dimensional space of environmental aspects, even including non-physical facets like emotions. Hence, models for representing context have evolved from using simple key-value pairs to using current methods and techniques derived from artificial intelligence and knowledge management, e.g., logic, object relationship models, and ontologies.

Appropriate context management methods are an important prerequisite for using this contextual information, e.g., to determine or assign a context to a situation, to cope with the fuzziness of context information and, especially because of mobility, to deal with rapidly changing environments and unstable information sources. Therefore, advanced models, methods, and tools are needed to provide mechanisms and techniques for structured storage of contextual information, to provide effective ways to retrieve it, and to enable integration of context and application knowledge. This entails the need for artificial intelligence mechanisms in context-aware applications.

Nine papers from the Second International Workshop on Modeling and Retrieval of Context, MRC 2005, held at the 19th International Joint Conference on Artificial Intelligence, in Edinburgh, Scotland, July 31–August 1, 2005, were selected and extended for this book. A major goal of the workshop was to study, understand, and explore the handling of context in IT applications. The following papers illustrate the state of the art of context modeling and elicitation as well as identification and application of context in different application scenarios.

Anders Kofod-Petersen and Jörg Cassens propose an interdisciplinary approach, using Activity Theory, to model context and then populate the model for assessing situations in a pervasive computing environment. Thus, they provide a knowledge-intensive context model from a socio-technical perspective.

Sven Schwarz introduces a context model for personalized knowledge management applications. The approach is based on an ontology that describes different aspects of a knowledge worker's information needs. Contextual information is gathered by user observation. The paper describes several implemented example applications.

Dominik Heckmann describes a service for modeling situations and retrieving contextual information in mobile and ubiquitous computing environments using Semantic Web technology. The paper introduces a General User Model and Context Ontology, called GUMO.

The next three papers describe layered architectures for context management, each with a different focus. The first paper describes a generic architecture, the second paper focusses on contextualized decisions, and the third paper deals with imperfection and aging of contextual information.

Kayu Wan, Vasu Alagar, and Joey Paquet describe and evaluate a generic, component-based architecture for developing context-aware systems. They propose a three tier model in which the first tier deals with perception, the second tier deals with context management, and the third tier deals with context adaptation.

The paper by Oana Bucur, Philippe Beaune, and Olivier Boissier discusses steps towards contextualized decisions. It addresses the problem of how to distinguish relevant from non-relevant context for a given task. The basis of the solution is a context definition and model for a context-aware agent that is able to learn to select relevant contexts. The three tier architecture comprises a layer of context sources, a context management layer, and a layer of agents that reason with context.

Andreas Schmidt deals with imperfection handling and controlled aging of contextual information. The paper presents a three layer model and its application to a context-aware learning environment for corporate learning support. The model distinguishes an internal layer with a context information base, a logical layer containing context feature values, and an external layer where the context information is applied.

The papers by Maria Chantzara and Miltiades Anagostou and by Aviv Segev (as well as by Bucur, Beaune, and Boissier mentioned above), present approaches to identifying appropriate contexts.

Maria Chantzara and Miltiades Anagnostou look into quality-aware discovery of context information. They introduce the Context Matching Engine which allows discovery of appropriate context sources for customized context-aware services.

The paper by Aviv Segev deals with identification of multiple contexts of a situation. The context recognition algorithm presented uses the Internet as a knowledge base. In the example described the real-time approach is successfully applied to ongoing textual conversations such as chats.

The paper by Michael Fahrmair, Wassiou Sitou, and Bernd Spanfelner proposes a generic mechanism for designing context awareness and adaptation behavior with formal methods. It introduces the concept of an adaptation context as a characterization of the system after carrying out an adaptation.

We hope that this research snapshot will be a useful foundation for future research on modeling and reasoning about context.

February 2006   Thomas R. Roth-Berghofer
                                                                        Stefan Schulz
                                                                        David B. Leake

Organization

Workshop Chairs

Thomas R. Roth-Berghofer (DFKI GmbH, Germany)
Stefan Schulz (The e-Spirit Company GmbH, Germany)
David B. Leake (Indiana University, USA)

Program Committee

Agnar Aamodt (Norwegian University of Science and Technology)
Paolo Bouquet (University of Trento, Italy)
Shannon Bradshaw (University of Iowa, USA)
Patrick Brezillon (University of Paris, France)
Hans-Dieter Burkhard (Humboldt University Berlin, Germany)
Andreas Dengel (DFKI GmbH, Kaiserslautern, Germany)
Anind Dey (University of California, Berkeley, USA)
Babak Esfandiari (Carleton University, Canada)
Mehmet H. Göker (PricewaterhouseCoopers, USA)
Avelino Gonzalez (University of Central Florida, USA)
Theo G. Kanter (Ericsson Research, Sweden)
Jeroen Keppens (King's College London, UK)
Mohamed Khedr (Arab Academy for Science and Technology, Egypt)
Thomas Kunz (Carleton University, Canada)
Kristof Van Laerhoven (Lancaster University, UK)
Ana G. Maguitman (Indiana University, USA)
Heiko Maus (DFKI GmbH, Kaiserslautern, Germany)
Enrico Rukzio (University of Munich, Germany)
Thomas Strang (DLR, Germany)

Table of Contents

Using Activity Theory to Model
Context Awareness

Anders Kofod-Petersen and Jörg Cassens

Department of Computer and Information Science (IDI),
Norwegian University of Science and Technology (NTNU),
7491 Trondheim, Norway
{anderpe,cassens}@idi.ntnu.no
http://www.idi.ntnu.no/

Abstract. One of the cornerstones of any intelligent entity is the ability to understand how occurrences in the surrounding world influence its own behaviour. Different states, or situations, in its environment should be taken into account when reasoning or acting. When dealing with different situations, context is the key element used to infer possible actions and information needs. The activities of the perceiving agent and other entities are arguably one of the most important features of a situation; this is equally true whether the agent is artificial or not.

This work proposes the use of Activity Theory to first model context and further on populate the model for assessing situations in a pervasive computing environment. Through the socio-technical perspective given by Activity Theory, the knowledge intensive context model, utilised in our ambient intelligent system, is designed.

1 Introduction

The original vision of ubiquitous computing proposed by Weiser [1] envisioned a world of simple electronic artefacts, which could assist users in their day to day activities. This vision has grown significantly. Today the world of ubiquitous computing, pervasive computing or ambient intelligence uses visions and scenarios that are far more complex. Many of the scenarios of today envision pro-active and intelligent environments, which are capable of making assumptions and selections on their own accord.

Several examples exist in the contemporary literature, such as the help Fred receives from the omnipresent system Aura in [2, p. 3], and the *automagic* way that Maria gets help on her business trip in [3, p. 4]. More examples and comments can be found in [4]. Common to many of these examples are the degree of autonomy, common sense reasoning, and situation understanding the systems involved exhibit.

To be truly pro-active and be able to display even a simple level of common sense reasoning, an entity must be able to appreciate the environment which it inhabits; or to understand the situations that occur around it. When humans interpret situations, the concept of context becomes important. Humans use an

T.R. Roth-Berghofer, S. Schulz, and D.B. Leake (Eds.): MRC 2005, LNAI 3946, pp. 1–17, 2006.

abundance of more or less subtle cues as context and thereby understand, or at least assess, situations. The ability to acquire context and thereby fashion an understanding of situations, is equally important for artefacts that wish to interact (intelligently) with the real world. Systems displaying this ability to acquire and react to context are known as *context-aware* systems.

A major shortfall of the research into context-aware systems is the lack of a common understanding of what a context model is, and perhaps more importantly, what it is not. This shortfall is very natural, since this lack of an agreed definition of context also plagues the real world. No common understanding of what context is and how it is used exists. So, it is hardly surprising that it is hard to agree on the artificial world that IT systems represent.

Most of the research today has been focused on the technical issues associated with context, and the syntactic relationships between different concepts. Not so much attention has been given to context from a knowledge level [5] perspective or an analysis of context on the level of socio-technical systems [6].

This is the main reason for the approach chosen here. It should be feasible to look at how we can use socio-technical theories to design context-aware systems to supply better services to the user, in a flexible and manageable way. The approach should facilitate modelling at the knowledge level as well and furthermore enable the integration of different knowledge sources and the presentation of knowledge content to the user.

It can be stated that one of the most important context parameters available in many situations is the *activity* performed by an entity present in the environment. We therefore believe that by focusing on activities we will gain a better understanding of context and context awareness; thus bringing us closer to realise truly ambient intelligent systems.

Several approaches to examine activity have been proposed, like e.g. Actor-Network Theory [7], Situated Action [8] or the Locales Framework [9]. One of the most intriguing theories, however, is Activity Theory based on the works of Vygotsky and Leont'ev [10, 11, 12]. This work proposes the use of Activity Theory to model context and to describe situations.

Although our approach is general, in the sense that it is applicable to different domains, we are not trying to define a context model which will empower the system to be universally context aware, meaning it will be able to build its own context model on the fly. Although this would be a prerequisite for truly intelligent systems, IT-systems are usually designed for specific purposes and with specific tasks in mind where the system has to support human users. They are used by people with specific needs and qualifications, and should preferably adapt to changes in these needs over time [13, 14]. The aim of the work presented in this article is to assist the design of such systems which are tailored to support such kind of human work.

This article is organised as follows: first some background work on the use of context in cognition is covered. Secondly, some important concepts of Activity Theory are introduced. This is followed by an explanation of how Activity Theory can be utilised to model contextual information, including an illustrative example. In Section 5, the knowledge model, including context employed in this work, is described. Finally, some pointers for future work are presented.

2   Context in Cognition

The concept of context is closely related to reasoning and cognition in humans. Even though context might be important for reasoning in other animals, it is common knowledge that context is of huge importance in human reasoning.

Beside the more mechanistic view on reasoning advocated by neuroscience, psychology and philosophy play important roles in understanding human cognition. It might not be obvious how computer science is related to knowledge about human cognition. However, many sub-fields in computer science are influenced by our knowledge about humans; and other animals.

The field of Artificial Intelligence has the most obvious relations to the study of reasoning in the real world, most prominently psychology and philosophy. Since AI and psychology are very closely related and context is an important aspect of human reasoning, context also plays an important role in the understanding and implementation of Artificial Intelligence.

AI has historically been closely connected to formal logic. Formal logic is concerned with explicit representation of knowledge. This leads to the need to codify all facts that could be of importance. This strict view on objective truth is also known in certain directions within philosophy, where such a concept of knowledge as an objective truth exists. This can be traced back to e.g. the logic of Aristotle who believed that some subset of knowledge had that characteristic (Episteme). This view stands in stark contrast to the views advocated by people such as Polanyi, who argues that no such objective truth exists and all knowledge is at some point personal and hidden (tacit) [15].

Since context is an elusive type of knowledge, where it is hard to quantify what type of knowledge is useful in a certain situation, and possibly why, it is obvious that it does not fit very well with the strict logical view on how to model the world. Ekbia and Maguitman [16] argue that this has led to the fact that context has largely been ignored by the AI community. This observation still holds some truth, despite some earlier work on context and AI, like Doug Lenat's discussion of context dimensions [17], and the other work we discuss later in this section.

Ekbia and Maguitman's paper is not a recipe on how to incorporate contextual reasoning into logistic systems, but rather an attempt to point out the deficiencies and suggest possible directions AI could take to include context. Their work builds on the work by the American philosopher John Dewey. According to Ekbia and Maguitman, Dewey distinguishes between two main categories of context: spatial and temporal context, coherently know as background context; and selective interest. The spatial context covers all contemporary parameters. The temporal context consists of both intellectual and existential context. The intellectual context is what we would normally label as background knowledge, such as tradition, mental habits, and science. Existential context is combined with the selective interest related to the notion of situation. A situation is in this work viewed as a confused, obscure, and conflicting thing, where a human reasoner attempts to make sense of this through the use of context. This view, by Dewey, on human context leads to the following suggestion by the pragmatic approach [16, p. 5]:

1. Context, most often, is not explicitly identifiable.
2. There are no sharp boundaries among contexts.
3. The logical aspects of thinking cannot be isolated from material considerations.
4. Behaviour and context are jointly recognisable.

Once these premises have been set, the authors show that the logical approach to (artificial) reasoning has not dealt with context in any consistent way. The underlying argument is that AI has been using an absolute separation between mind and nature, thus leading to the problems associated with the use of context. This view on the inseparability of mind and nature is also based on Dewey's work. This view is not unique for Dewey. In recent years this view has been proposed in robotics as *situatedness* by Brooks [18, 19, 20], and in ecological psychology by J. J. Gibson [21].

Through the discussion of different logic-based AI methods and systems, the authors argue that AI has not yet parted company with the limitations of logic with regards to context. Furthermore, they stress the point of intelligence being action-oriented; based on the notion of situations described above.

The notion of intelligence being action-oriented, thus making context a tool for selecting the correct action, is shared by many people within the computer science milieu. Most notably the work by Strat [22], where context is applied to select the most suitable algorithm for recognition in computer vision, and by Öztürk and Aamodt [23] who utilised context to improve the quality and efficiency of Case-Based Reasoning.

Strat [22] reports on the work done in computer vision to use contextual information in guiding the selection of algorithms in image understanding. When humans observe a scene they utilise a large amount of information (context) not captured in the particular image. At the same time, all image understanding algorithms use some assumptions in order to function, creating an epistemic bias. Examples are algorithms that only work on binary images, or that are not able to handle occlusions.

Strat defines three main categories of context: *physical*, being general information about the visual world independent of the conditions under which the image was taken; *photogrammetric*, which is the information related the acquisition of the image; and *computational*, being information about the internal state of the processing. The main idea in this work is to use context to guide the selection of the image-processing algorithms to use on particular images. This is very much in line with the ideas proposed by Ekbia and Maguitman, where intelligence is action-oriented, and context can be used to bring order to diffuse and unclear situations.

This action-orientated view on reasoning and use of context is also advocated by Öztürk and Aamodt [23]. They argue that the essential aspects of context are the notion of *relevance* and *focus*. To facilitate improvements to Case-Based Reasoning a context model is constructed. This model builds on the work by Hewitt, where the notion of *intrinsic* and *extrinsic* context types are central. According to Hewitt, intrinsic context is information related to the target item in a reasoning process, and extrinsic is the information not directly related to the target item. This distinction is closely related to the concepts of *selective interest* and *background context* as described by Dewey. The authors refine this view by focusing on the intertwined relationship between the *agent* doing the reasoning, and the *characteristics* of the problem to be solved. This is exactly the approach recognised as being missing in AI by Ekbia and Maguitman.

Öztürk and Aamodt build a taxonomy of context categories based on this merger of the two different worlds of information (internal vs. external). Beside this categorisation, the authors impose the action, or task, oriented view on knowledge in general, and contextual knowledge in particular. The goal of an agent *focuses* the attention, and thereby the knowledge needed to execute tasks associated with the goal. The example domain in their paper is from medical diagnostics, where a physician attempts to diagnose a patient by the hypothesise-and-test strategy. The particular method of diagnostics in this Case-Based Reasoning system is related to the strategy used by Strat. They differ insofar that Strat used contextual information to select the algorithms to be used, whereas Öztürk and Aamodt have, prior to run-time, defined the main structure of a diagnostic situation, and only use context to guide the sub-tasks in this process.

Zibetti et al. [24] focus on the problem of how agents understand situations based on the information they can perceive. To our knowledge, this work is the only one that does not attempt to build an explicit ontology on contextual information prior to run-time. The idea is to build a (subjective) taxonomy of ever-complex situations solely based on what a particular agent gathers from the environment in general, and the behaviour of other agents in particular.

The implementation used to exemplify this approach contains a number of agents "living" in a two-dimensional world, where they try to make sense of the world by assessing the spatial changes to the environment. Obviously the acquisition of knowledge starting with a *tabula rasa* is a long and tedious task for any entity. To speed up the process the authors predefined some categories with which the system is instantiated.

All in all, this approach lies in between a complete bottom-up and the top-down approaches described earlier.

3   Activity Theory

In this section, we concentrate on the use of Activity Theory (AT) to support the modelling of context. Our aim is to use AT to analyse the use of technical artefacts as instruments for achieving a predefined goal in the work process as well as the role of social components, like the division of labour and community rules. This helps us to understand what pieces of knowledge are involved and the social and technological context used when solving a given problem.

First in this section, we will give a short summary of aspects of AT that are important for this work. See [25] for a short introduction to AT and [26, 27] for deeper coverage. The theoretical foundations of AT in general can be found in the works of Vygotsky and Leont'ev [10, 11, 12].

Activity Theory is a descriptive tool to help understand the unity of consciousness and activity. Its focus lies on individual and collective work practise. One of its strengths is the ability to identify the role of material artefacts in the work process. An activity (Fig. 1) is composed of a subject, an object, and a mediating artefact or tool. A subject is a person or a group engaged in an activity. An object is held by the subject, and the subject has a goal directed towards the object he wants to achieve, motivating the activity and giving it a specific direction.

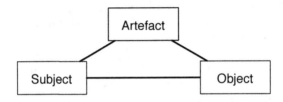

Fig. 1. Activity Theory: The basic triangle of Mediation

Some basic properties of Activity Theory are:

- **Hierarchical structure of activity:** Activities (the topmost category) are composed of goal-directed actions. These actions are performed consciously. Actions, in turn, consist of non-conscious operations.
- **Object-orientedness:** Objective and socially or culturally defined properties. Our way of doing work is grounded in a praxis which is shared by our co-workers and determined by tradition. The way an artefact is used and the division of labour influences the design. Hence, artefacts pass on the specific praxis they are designed for.
- **Mediation:** Human activity is mediated by tools, language, etc. The artefacts as such are not the object of our activities, but appear already as socio-cultural entities.
- **Continuous Development:** Both the tools used and the activity itself are constantly reshaped. Tools reflects accumulated social knowledge, hence they transport social history back into the activity and to the user.
- **Distinction between internal and external activities:** Traditional cognitive psychology focuses on what is denoted internal activities in Activity Theory, but it is emphasized that these mental processes cannot be properly understood when separated from external activities, that is the interaction with the outside world.

Taking a closer look on the hierarchical structure of activity, we can find the following levels:

- **Activity:** An individual activity is for example to check into a hotel, or to travel to another city to participate at a conference. Individual activities can be part of collective activities, e.g. when someone organises a workshop with some co-workers.
- **Actions:** Activities consist of a collections of actions. An action is performed consciously, the hotel check-in, for example, consists of actions like presenting the reservation, confirmation of room types, and handover of keys.
- **Operations:** Actions consist themselves of collections of non-conscious operations. To stay with our hotel example, writing your name on a sheet of paper or taking the keys are operations. That operations happen non-consciously does not mean that they are not accessible.

It is important to note that this hierarchical composition is not fixed over time. If an action fails, the operations comprising the action can get conceptualised, they become conscious operations and might become actions in the next attempt to reach the overall goal. This is referred to as a breakdown situation. In the same manner, actions can become automated when done many times and thus become operations. In this way, we gain the ability to model a change over time.

An expanded model of Activity Theory, Cultural Historical Activity Theory (CHAT), covers the fact that human work is done in a social and cultural context (compare e.g. [28, 29]). The expanded model (depicted in Fig. 2) takes this aspect into account by adding a *community* component and other mediators, especially *rules* (an accumulation of knowledge about how to do something) and the *division of labour*.

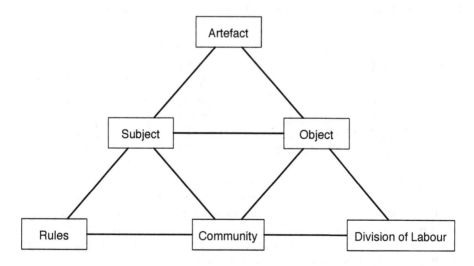

Fig. 2. Cultural Historical Activity Theory: Expanded triangle, incorporating the community and other mediators

In order to be able to model that several subjects can share the same object, we add the community to represent that a subject is embedded in a social context. Now we have relationships between subject and community and between object and community, respectively. These relationships are themselves mediated, with rules regarding to the subject and the division of labour regarding to the object.

This expanded model of AT is the starting point for our use of AT in the modelling of context for intelligent systems.

4   Activity Theory and Context Awareness

The next step is to identify which aspects of an Activity Theory based analysis can help us to capture a knowledge level view of contextual knowledge that should be incorporated into an intelligent system. This contextual knowledge should include knowledge about the acting subjects, the objects towards which activities are directed and the community as well as knowledge about the mediating components, like rules or tools.

4.1   Activity Theory for the Identification of Context Components

As an example, we want the contextual knowledge to contain both information about the acting subject itself (like the weight or size) and the tools (like a particular software used in a software development process). To this end, we propose a mapping from the basic structure of an activity into a taxonomy of contextual knowledge as depicted in Table 1 (the taxonomy is described in more detail in Section 5). We can see that the personal context contains information we would associate with the acting subject itself.

Table 1. Basic aspects of an activity and their relation to a taxonomy of contextual knowledge

CHAT aspect	Category
Subject	Personal Context
Object	Task Context
Community	Spatio-Temporal Context
Mediating Artefact	Environmental Context
Mediating Rules	Task Context
Mediating Division of Labour	Social Context

We would like to point out that we do not think that a strict one to one mapping exists or is desirable at all. Our view on contextual knowledge is contextualised itself in the sense that different interpretations exist, and what is to be considered contextual information in one setting is part of the general knowledge model in another one. Likewise, the same piece of knowledge can be part of different categories based on the task at hand.

The same holds for the AT based analysis itself: the same thing can be an object and a mediating artefact from different perspectives and in different task settings. The mapping suggested here should lead the development process and allow the designer to focus on knowledge-level aspects instead of being lost in the modelling of details without being able to see the relationship between different aspects on a socio-technical system level.

As an example, let us consider a software development setting where a team is programming a piece of software for a client. The members of the team are all *subjects* in the development process. They form a *community* together with representatives of the client and other stake-holders. Each member of the team and personnel from other divisions of the software company work together in a *division of labour*. The *object* at hand is the unfinished prototype, which has to be transformed into something that can be handed over to the client. The task is governed by a set of *rules*, some explicit like coding standards some implicit like what is often referred to as a working culture. The programmers use a set of *mediating artefacts* (tools), like methods for analysis and design, programming tools, and documentation.

When we design a context-aware system for the support of this task, we include information about the user of the system (*subject*) in the *personal context* and about the other team members in the *environmental context*. Aspects regarding the special application a developer is working on (*objects*) are part of the *task context*, it will change when the same user engages in a different task (lets say he is looking for a restaurant). The *rules* are part of the *task context* since they are closely related to the task at hand – coding standards will not be helpful when trying to find a restaurant. We find the tool aspects (*artefacts*) in the *environmental context* since access to the different tools is important for the ability of the user to use them. Knowledge about his co-workers and other stake-holders (*community*) are modelled in the *spatio-temporal context*. Finally, his interaction with other team members (*division of labour*) is described as part of the *social context*.

In the design process, we can also make use of the hierarchical structure of activities. On the topmost level, we can identify the *activities* the context-aware system should support. By this, we can restrict the world view of the system and make the task of developing a context model manageable. Further on, we can make use of the notion of *actions* to identify the different situations the system can encounter. This helps us to asses the different knowledge sources and artefacts involved in different contexts, thereby guiding the knowledge acquisition task. Finally, since *operations* are performed subconsciously, we get hints on which processes should be supported by automatic and proactive behaviour of the system.

Let us consider our example again. We know that the *activity* we want to support is the development of an IT system. Therefore, we can restrict ourselves to facets of the world which are related to the design process, and we do not (necessarily) have to take care of supporting e.g. meetings some of the team members have as players at the company's football team. On the other hand, the system has to be concerned with meetings with the customer. Further on, different *actions* which are also part of the activity should be supported, like e.g. team meetings or programming sessions, and the different *actions* involved can lead to the definition of different situations or contexts.

A context-aware application should therefore at all times know in which *action* the user is engaged. This is, in fact, the main aspect of our understanding of the term *context awareness*. At last, to support the *operations* of the user, it might be necessary to proactively query different knowledge sources or request other resources the user might need without being explicitly told to do so by the user. This is at the core of what we refer to as *context sensitivity* in order to distinguish between these two different aspects of context.

It is important to keep in mind that the hierarchical structure of activities is in a constant state of flux. Activity Theory is also capable of capturing changing contexts in break-down situations. Lets consider that a tool used in the development process, such as a compiler, stops working. The operation of evoking the compiler now becomes a conscious action for the debugging process. The focus of the developer shifts away from the client software to the compiler. He will now be involved in a different task where he probably will have to work together with the system administrators of his work-station. In this sense other aspects of the activity, such as the community, change as well. It is clear that the contextual model should reflect these changes. The ability of Activity Theory to identify possible break-down situations makes it possible for the system designer to identify these possible shifts in situation and model the anticipated behaviour of the system.

4.2   Other Aspects of Activity Theory and Context

Other work on the use of AT in modelling context has been conducted e.g. by Kaenampornpan and O'Neill [30]. This work is focusing on modelling features of the world according to an activity theoretic model. However, the authors do not carry out a knowledge level analysis of the activities. We argue that our knowledge intensive approach has the advantage of giving the system the ability to reason about context so that it does not have to rely on pattern matching only. This is helpful especially in situations where not all the necessary features are accessible by the system, for example because of limits of sensory input in mobile applications. On the other hand, Kaenampornpan and O'Neill further on develop a notion of history of context in order to elicit a users goals [31]. This work deals with the interesting problem of representing the user's history in context models which we have not addressed explicitly in this article.

Li and Landay [32] propose an activity based design tool for context aware applications. The authors' focus lies not on the use of Activity Theory in the context model itself but on supporting the designer of context aware applications with a rapid-prototyping tool. An interesting idea is the proposed integration of temporal probabilistic models.

Wiberg and Olsson [33] make also use of Activity Theory, but their focus lies on the design of context aware tangible artefacts. The usage situation is well defined upfront and no reasoning about the context has to be done.

When we look at the design of IT-systems in general and not only the issue of context-awareness, we find that Activity Theory has been applied to many different areas of system development. For example, AT was used in health care settings as a tool to support development of information systems [34]. It has also been used in the design of augmented reality systems, as reported in [35] and for a posteriori analysis of computer systems in use [36]. A comparative survey of five different AT based methods for information systems development with pointers to additional examples was conducted by Queak and Shah [37].

In our own work, we are also using Activity Theory to support modelling other, not context depending aspects of intelligent systems. For example are we focusing on breakdown situations in order to enhance the explanatory capabilities of knowledge-rich Case-Based Reasoning systems [38].

5   Context Model

The context model used in this work draws on a subjective view on situations. That is, even though the model is general, any instance of the model belongs to one user only. Thus, as in [24], any situation will be described form the personal perspective, leading to the possibility of many instances describing the "same" situation. This is in contrast to the leading perspective, where a system will describe *objective* situations, and leans towards Polanyi's perspective of all knowledge being personal [15].

In the extreme consequence the model used by any subject could also be personal and unique. However, to avoid the problem of a *tabula rasa* we have chosen a pragmatical view on how to model context. The model is based on the definition of context given by Dey [39], applying the following definition:

> Context is the set of suitable environmental states and settings concerning a user, which are relevant for a situation sensitive application in the process of adapting the services and information offered to the user.

This definition from Dey does not explicitly state that context is viewed as knowledge. However, we believe that the knowledge intensive approach is required if we wish a system to display many of the characteristics mentioned in the introduction. At the same time we also adhere to the view advocated by Brézillon and Pomerol [40] that context is not a special kind of knowledge. They argue that context is in the eye of the beholder: "... knowledge that can be qualified as 'contextual' depends on the context!" [40, p.7]

Even though we argue for a context model where context is not a special type of information, we also believe that only a pragmatical view on context will enable us to construct actually working systems. Following this pragmatic view we impose a taxonomy on the context model in the design phase (see Fig. 4). This taxonomy is inherited from the context-aware tradition and adopted to make use of the general concepts we find in Activity Theory.

The context is divided into five sub-categories (a more thorough discussion can be found in [41] or [42]):

1. **Environmental context:** This part captures the users surroundings, such as things, services, people, and information accessed by the user.

2. **Personal context:** This part describes the mental and physical information about the user, such as mood, expertise and disabilities.
3. **Social context:** This describes the social aspects of the user, such as information about the different roles a user can assume.
4. **Task context:** the task context describe what the user is doing, it can describe the user's goals, tasks and activities.
5. **Spatio-temporal context:** This type of context is concerned with attributes like: time, location and the community present.

The model depicted in Fig. 4 shows the top-level ontology. To enable the reasoning in the system this top-level structure is integrated with a more general domain ontology, which describes concepts of the domain (*e.g.*, Operating Theatre, Ward, Nurse, Journal) as well as more generic concepts (Task, Goal, Action, Physical Object) in a multi-relational semantic network. The model enables the system to infer relationships between concepts by constructing context-dependent paths between them. We are approaching the situation assessment by applying knowledge-intensive Case-Based Reasoning [43]. One of the important aspects of knowledge-intensive Case-Based Reasoning is the ability to match two

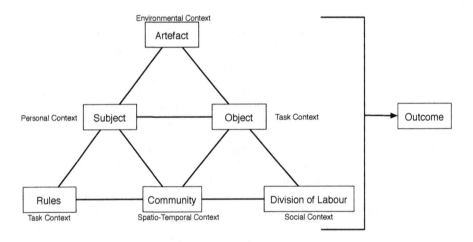

Fig. 3. Mapping from Activity Theory to context model

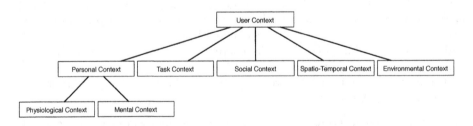

Fig. 4. Context taxonomy

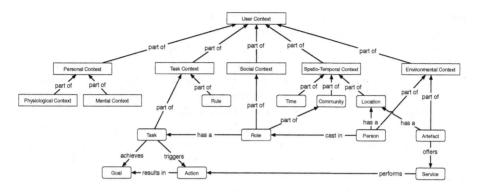

Fig. 5. Populated context structure

case features that are syntactically different, by explaining why they are similar [44, 45].

The generic concepts are partly gathered through the use of activity theoretic analysis. These concepts include the six aspects shown in Fig. 3. The top-level taxonomy including the concepts acquired from AT is depicted in Fig. 5. The context model is now primed to model situations and the activities occurring within them.

If we look at the model we can see how each of the AT aspects are modelled. The *artefact* exists within the *environmental context*, where it can offer *services* that can perform *actions*, which assist the *subject* (described in the *personal context*) in achieving the *goals* of the *role* (in the *social context*) played by the subject. Other *persons*, being part of the situation through the environmental context, can also affect the *outcome* (goal) of the situation. They are cast in different roles that are part of the *community* existing in the *spatio-temporal context*. The roles also implicitly define the *division of labour* in the community. The *rules* governing the subject are found in the *task context*.

6   Ongoing and Future Work

We have outlined how the design of context-aware systems can benefit from an analysis of the underlying socio-technical system. We have introduced a knowledge-level perspective on the modelling task, which makes it possible to identify aspects of knowledge that should be modelled into the system in order to support the user with contextual information. We have furthermore proposed a first mapping from an Activity Theory based analysis to different knowledge components of a context model. The basic aspects of our socio-technical model fit nicely to the taxonomy of context categories we have introduced before, thus making AT a prime candidate for further research.

The use of Activity Theory allows for system designers to develop the general models of activities and situations. General models are necessary to support the initial usage of the system. They are an important prerequisite for the Case-Based Reasoning system to integrate new situations; thereby adapting to the personal and subjective perspective of the individual user.

In Section 5 we have formulated the problem of identifying the tasks connected to a particular situation, the goals of the user, and the artefacts and information sources used. We argue that our Activity Theory based approach is capable of integrating these cognitive aspects into the modelling process.

The integration of an *a posteriori* method of analysis with design methodologies is always challenging. One advantage AT has is that it is process oriented, which corresponds to a view on systems design where the deployed system itself is not static and where the system is able to incorporate new knowledge over time [46]. Activity Theory has its blind spots, such as modelling the user interaction of the interface level. However, in this particular work we are not focusing on user interfaces; thus, these deficiencies do not affect this work directly. Still, one of our future goal is to combine AT with other theories into a framework of different methods supporting the systems design process [47].

Nevertheless, one of the next steps is to formalise the relationship between different elements of an AT based analysis and the knowledge contained in the different contextual aspects of our model. This more formalised relationship is being put to the test on a context modelling task, using an AT based analysis of a socio-technical system to support the design of our context-aware intelligent system (see for an example [48] for a description of the system).

We have recently initiated a project where everyday situations in a health care setting are being observed and documented. These observations are being used to test the situation assessment capabilities of our system. We have used a modelling approach based on Cultural Historical Activity Theory. This allows us to identify the different actions the medical staff is involved with and the artefacts and information sources used.

We have already instantiated a context model for this scenario using the topology described earlier in this paper. We are currently in the process of populating the model based on our observations. At the same time, we are refining our knowledge engineering methodologies for translating the findings into a knowledge model.

Our system also includes an agency part, which is described in [49]. Based on the context-aware situation assessment being carried out, this agency supplies context-sensitive problem solving [50]. We are in the process of extending the analysis of the situations to model the way our decompose agent decomposes and solves problems.

Acknowledgements

Part of this work was carried out in the AmbieSense project, which was supported by the EU commission (IST-2001-34244).

References

1. Weiser, M.: The computer for the 21st century. Scientific American (1991) 94–104
2. Satyanarayanan, M.: Pervasive computing: Vision and challenges. IEEE Personal Communications **8** (2001) 10–17
3. Ducatel, K., Bogdanowicz, M., Scapolo, F., Leijten, J., Burgelman, J.C.: ISTAG Scenarios for Ambient Intelligence in 2010. Technical report, IST Advisory Group (2001)
4. Lueg, C.: Representaion in Pervasive Computing. In: Proceedings of the Inaugural Asia Pacific Forum on Pervasive Computing. (2002)
5. Newell, A.: The Knowledge Level. Artificial Intelligence **18** (1982) 87–127
6. Lueg, C.: Looking under the rug: On context-aware artifacts and socially adept technologies. In: Proceedings of the Workshop The Philosophy and Design of Socially Adept Technologies at the ACM SIGCHI Conference on Human Factors in Computing Systems (CHI 2002), National Research Council Canada NRC 44918 (2002)
7. Latour, B.: Science in Action: How to Follow Scientists and Engineers Through Society. Harvard University Press (1988)
8. Suchman, L.A.: Plans and Situated Actions: The Problem of Human-Machine Communication. Cambridge University Press, New York, NY, USA (1987)
9. Fitzpatrick, G.: The Locales Framework: Understanding and Designing for Cooperative Work. Ph.d. thesis, The University of Queensland, Australia (1998)
10. Vygotski, L.S.: Mind in Society. Harvard University Press, Cambridge, MA (1978)
11. Vygotski, L.S.: Ausgewählte Schriften Bd. 1: Arbeiten zu theoretischen und methodologischen Problemen der Psychologie. Pahl-Rugenstein, Köln (1985)
12. Leont'ev, A.N.: Activity, Consciousness, and Personality. Prentice-Hall (1978)
13. McSherry, D.: Mixed-Initative Dialogue in Case-Based Reasoning. In: Workshop Proceedings ECCBR-2002, Aberdeen (2002)
14. Totterdell, P., Rautenbach, P.: Adaptation as a Design Problem. In: Adaptive User Interfaces. Academic Pres (1990) 59–84
15. Polanyi, M.: Personal Knowledge: Towards a Post-critical Philosophy. N.Y.: Harper & Row (1964)
16. Ekbia, H.R., Maguitman, A.G.: Context and relevance: A pragmatic approach. Lecture Notes in Computer Science **2116** (2001) 156–169
17. Lenat, D.: The Dimensions of Context-Space. Technical report, Cycorp, Austin, TX (1998)
18. Brooks, R.A.: Planning is Just a Way of Avoiding Figuring Out What to Do Next. Technical report, MIT (1987)
19. Brooks, R.A.: Intelligence without Representation. Artificial Intelligence **47** (1991) 139–159
20. Brooks, R.A.: New Approaches to Robotics. Science **253** (1991) 1227–1232
21. Gibson, J.J.: The Ecological Approach to Visual Perception. Houghton Mifflin (1979)
22. Strat, T.M.: Employing contextual information in computer vision. In: DARPA93. (1993) 217–229
23. Öztürk, P., Aamodt, A.: A context model for knowledge-intensive case-based reasoning. International Journal of Human Computer Studies **48** (1998) 331–355
24. Zibetti, E., Quera, V., Beltran, F.S., Tijus, C.: Contextual categorization: A mechanism linking perception and knowledge in modeling and simulating perceived events as actions. Modeling and using context (2001) 395–408

25. Nardi, B.A.: A Brief Introduction to Activity Theory. KI – Künstliche Intelligenz (2003) 35–36
26. Bødker, S.: Activity theory as a challenge to systems design. In Nissen, H.E., Klein, H., Hirschheim, R., eds.: Information Systems Research: Contemporary Approaches and Emergent Traditions. North Holland (1991) 551–564
27. Nardi, B.A., ed.: Context and Consciousness. MIT Press, Cambridge, MA (1996)
28. Kutti, K.: Activity Theory as a Potential Framework for Human-Computer Interaction Research. [27] 17–44
29. Mwanza, D.: Mind the Gap: Activity Theory and Design. Technical Report KMI-TR-95, Knowledge Media Institute, The Open University, Milton Keynes (2000)
30. Kaenampornpan, M., O'Neill, E.: Modelling context: an activity theory approach. In Markopoulos, P., Eggen, B., Aarts, E., Croeley, J.L., eds.: Ambient Intelligence: Second European Symposium on Ambient Intelligence, EUSAI 2004. Volume 3295 of Lecture Notes in Computer Science., Springer Verlag (2004) 367–374
31. Kaenampornpan, M., O'Neill, E.: Integrating History and Activity Theory in Context Aware System Design. In: Proceedings of the 1st International Workshop on Exploiting Context Histories in Smart Environments (ECHISE). (2005)
32. Li, Y., Hong, J.I., Landay, J.A.: Topiary: a tool for prototyping location-enhanced applications. In: UIST '04: Proceedings of the 17th annual ACM symposium on User interface software and technology, New York, NY, USA, ACM Press (2004) 217–226
33. Wiberg, M., Olsson, C.: Designing artifacts for context awareness. In: Proceedings of IRIS 22 Enterprise Architectures for Virtual Organisations, Jyväskylä, Dept. of Computer Science and Information Systems, University of Jyväskylä (1999) 49–58
34. Korpela, M., Soriyan, H.A., Olufokunbi, K.C.: Activity analysis as a method for information systems development. Scandinavian Journal of Information Systems 12 (2001) 191–210
35. Fjeld, M., Lauche, K., Bichsel, M., Voorhoorst, F., Krueger, H., Rauterberg, M.: Physcial and Virtual Tools: Activity Theory Applied to the Design of Groupware. CSCW 11 (2002) 153–180
36. Bødker, S.: Applying Activity Theory to Video Analysis: How to Make Sense of Video Data in Human-Computer Interaction. [27]
37. Quek, A., Shah, H.: A Comparative Survey of Activity-based Methods for Information Systems Development. In Seruca, I., Filipe, J., Hammoudi, S., Cordeiro, J., eds.: Proceedings of 6th International Conference on Enterprise Information Systems (ICEIS 2004). Volume 5., Porto, Portugal (2004) 221–229
38. Cassens, J.: Knowing What to Explain and When. In Gervás, P., Gupta, K.M., eds.: Proceedings of the ECCBR 2004 Workshops. Number 142-04 in Technical Report of the Departamento de Sistemas Informáticos y Programación, Universidad Complutense de Madrid, Madrid (2004) 97–104
39. Dey, A.K.: Understanding and using context. Personal and Ubiquitous Computing 5 (2001) 4–7
40. Brézillon, P., Pomerol, J.C.: Contextual knowledge sharing and cooperation in intelligent assistant systems. Le Travail Humain 62 (1999) 223–246
41. Göker, A., Myrhaug, H.I.: User context and personalisation. In: Workshop proceedings for the 6th European Conference on Case Based Reasoning. (2002)
42. Kofod-Petersen, A., Mikalsen, M.: Context: Representation and Reasoning – Representing and Reasoning about Context in a Mobile Environment. Revue d'Intelligence Artificielle 19 (2005) 479–498

43. Aamodt, A.: Knowledge-intensive case-based reasoning in creek. In Funk, P., Calero, P.A.G., eds.: Advances in case-based reasoning, 7th European Conference, ECCBR 2004, Proceedings. (2004) 1–15
44. Aamodt, A.: Explanation-driven case-based reasoning. In Wess, S., Althoff, K., Richter, M., eds.: Topics in Case-based reasoning. Springer Verlag (1994) 274–288
45. Jære, M.D., Aamodt, A., Skalle, P.: Representing temporal knowledge for case-based prediction. In: Advances in case-based reasoning. 6th European Conference, ECCBR 2002, Lecture Notes in Artificial Intelligence, 2416, Springer Verlag (2002) 174–188
46. Aamodt, A.: Knowledge Acquisition and Learning by Experience – The Role of Case-Specific Knowledge. In Tecuci, G., Kodratoff, Y., eds.: Machine Learning and Knowledge Acquisition – Integrated Approaches. Academic Press (1995) 197–245
47. Cassens, J.: User Aspects of Explanation Aware CBR Systems. In Costabile, M.F., Paternó, F., eds.: Human-Computer Interaction – INTERACT 2005. Volume 3585 of LNCS., Rome, Springer (2005) 1087–1090
48. Kofod-Petersen, A., Mikalsen, M.: An Architecture Supporting implementation of Context-Aware Services. In Floréen, P., Lindén, G., Niklander, T., Raatikainen, K., eds.: Workshop on Context Awareness for Proactive Systems (CAPS 2005), Helsinki, Finland, HIIT Publications (2005) 31–42
49. Lech, T.C., Wienhofen, L.W.M.: AmbieAgents: A Scalable Infrastructure for Mobile and Context-Aware Information Services. In: AAMAS '05: Proceedings of the fourth international joint conference on Autonomous agents and multiagent systems, New York, NY, USA, ACM Press (2005) 625–631
50. Gundersen, O.E., Kofod-Petersen, A.: Multiagent Based Problem-solving in a Mobile Environment. In Coward, E., ed.: Norsk Informatikkonferance 2005, NIK 2005, Institutt for Informatikk, Universitetet i Bergen (2005) 7–18

A Context Model for Personal Knowledge Management Applications

Sven Schwarz

German Research Center for Artificial Intelligence (DFKI GmbH),
Kaiserslautern D-67608, Germany
Sven.Schwarz@dfki.de

Abstract. In the research project EPOS[1] we build a pro-active, context-sensitive support system to aid the user with his knowledge work, which is mostly about searching, reading, creating, and archiving of documents. In order to avoid distracting the user, the context gathering is realized by installable user observation plugins for standard applications such as Mozilla Firefox and Thunderbird.

The main part of this paper is about the definition of a context model for the personal knowledge management domain. The context model incorporates only contextual elements relevant to satisfy the knowledge worker's potential information need. It stores only information items known to the user (such as links to his own documents, folders, etc.), as well as, shared ontologies to assure an understanding of the context.

The context is modeled in RDF/S and can be retrieved by context-aware applications from the context support system via an XML-RPC call.

1   Introduction

Although the same term is used by linguists, psychologists, and computer scientists, "context" is understood in a variety of ways. To talk about context does not make sense without talking about its application and the scenario it is used in. We will therefore start with a short introduction of the research scenario.

The overall goal of the research project EPOS [4] (Evolving Personal to Organizational knowledge Spaces) is to build up *organizational/group-wide* structured knowledge from the information items and structures (already) present at the members of the organization/group. In order to encourage each worker to archive and structure his own knowledge, the worker himself should be the first person to utilize his own structuring effort. This is the knowledge management *bait*. Instead of doing it because of an obligation, the worker does it to enhance his own support, to make his own work faster and better.

Now, what does it mean to support a worker with his own structures and information items? There are two alternatives for potential support: First, a

---

[1] This work has been supported by a grant from The Federal Ministry of Education, Science, Research, and Technology (FKZ ITW–01 IW C01).

T.R. Roth-Berghofer, S. Schulz, and D.B. Leake (Eds.): MRC 2005, LNAI 3946, pp. 18–33, 2006.

sophisticated retrieval of structures (folders/categories) and information items (documents) will support the user when he searches for material. Second, according to the worker's current context, potentially relevant structures and information items can be presented in a pro-active way, i. e., without the worker even asking for them.

Without going into discussions about good, humane assistance interfaces, the second alternative demands for capturing the worker's context as unobtrusively as possible. Hence, we envision to realize a context-sensitive support being both, pro-active and unobtrusive, and this calls for an *automatic* supply of user actions to *automatically* capture the user's context. EPOS relies on user observation to gather evidences for contextual information, which raises privacy issues of course. However, the user can, of course, disable and enable the observation of his actions at any time. Tackling the privacy issue further is out this paper's scope.

The remainder of this paper is structured as follows: Section 2 describes our scenario and the knowledge worker's world. Section 3 defines an appropriate context model for that scenario. Section 4 then explains how the context model is being filled. Section 5 shows how the context can be retrieved and how it can used. Section 6 presents some concrete examples of applications already created using our context methodology. After some related work in section 7, the paper concludes with section 8.

2   The EPOS Scenario

A knowledge worker searches for documents, reads, writes, archives, and structures (classifies) them. A lot of the work is already done with the standard functionality of today's PCs, that is, by using the functionality of standard applications (web browser, email client, text processor) and the operating system (file system). Structuring/classification of documents is, for instance, typically done using file folders. However, classification using file folders poses two problems: First, a document can only be classified using one folder, and second, folder hier-

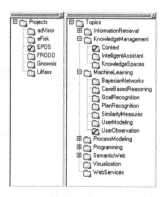

Fig. 1. Container hierarchies resemble the user's individual *views*, for example, relevant Projects and Topics

archies nearly never follow a consistent scheme. Therefore, EPOS initially starts with folder hierarchies already present on the user's PC. Beyond this, EPOS allows to build up additional, more consistent container hierarchies, which can then be used for a more consistent and multiple classification of documents.

A user will typically create several such container hierarchies, resembling his individual and subjective views of the world. Consequently, these container hierarchies are also called *views*. One view can be the *topics* view containing a hierarchy of containers representing scientific topics. To be more precise, such a view is a DAG (directed acyclic graph), so a topic can have more than just one super topic. That way the user creates a simple topic ontology. Analogously, the user can also build a *projects* view including containers for relevant projects. It can also be useful to have a *contacts* view, and maybe an *events* view, too, and so on. The point is, a document can be classified by putting it into *all* relevant containers. Let's take this paper as an example: It is about context and user observation, and it has to do with project EPOS. Hence, we can put that paper into the following three containers (see figure 1):

- `Topics/KnowledgeManagement/Context`
- `Topics/MachineLearning/UserObservation`
- `Projects/EPOS`

2.1   The Personal Knowledge Space

EPOS makes use of and accesses the *native structures* present on the user's PC. These are the documents (files), folders, emails, and so on. We pointed out, that EPOS also promotes building up multiple and more consistent container hierarchies (views). Documents can be put into these containers as a form of

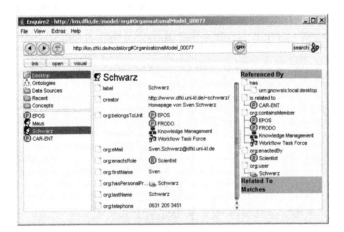

Fig. 2. EPOS allows for creating and browsing the personal semantic web. Example: The resource `Sven Schwarz` is an organizational person with some attributes and some links. Some attributed link (`dc:creator`) shows, he is the creator of his homepage.

classification. It is important to note, that for the personal knowledge space these view-containers do not simply *contain* some documents. Moreover, the *semantics* of the container is defined by the contained documents (instance-based semantics). That way, EPOS can help to classify both old and new documents.

Moreover, EPOS allows for relating several resources with each other. A document can be linked to some other document via a `dc:relation` link for example. Even more interesting is the usage of links with richer semantics: Using `dc:creator` some web page can be linked to some person (address book contact), declaring this person is the creator of that web page (see figure 2). To be more precise, EPOS supports building up a *personal semantic web* on the user's PC. This approach has been described in [11, 12].

2.2 The Knowledge Worker's World

Let's sum up the facts about our scenario by defining the (closed) world of a knowledge worker by giving a concrete list of the objects and tools he uses:

Document-like objects:

- Documents (HTML, plain text, wiki pages, PDF)
- Emails
- Notes (tiny text snippets created with EPOS Notes)

Objects useful for structuring/classification:

- File-Folders
- Topics (knowledge management, machine learning, etc.)
- Contacts, addresses (persons)
- Organizational entities (projects, organizations, etc.)
- Workflows and Tasks (from a workflow management system)

Note: All these structuring objects are potentially available both as individual ones (stored on the user's PC), as well as, organizational ones (hosted on some group-wide server) [5].

Applications:

- Text processor
- Web browser
- Email client
- Address book (of the email client)
- EPOS Notes (tiny note-taking tool, which is part of EPOS, see section 6.2)
- File system, brainFiler[2] (classification of documents)
- Organizational Repository (manages organizational entities, not handled in detail in this paper)
- FRODO TaskMan (Workflow Management System [6, 1])

Now, that we have explained the scenario and defined the (abstract) world of the knowledge worker, we can go over to explicitly model his context.

---

[2] http://www.brainbot.de

3   Modeling Context

When modeling the user's context we have to take into account his working (desktop) environment. In our scenario, a knowledge worker is mainly doing document oriented work (searching, reading, creating, archiving, classifying), as well as, organizational work (communicating with colleagues, executing workflow tasks). Hence, the user's context contains information about *currently or recently* read documents, relevant topics, relevant persons, relevant projects, etc., that is, exactly the objects listed in section 2.2. In EPOS, the user's context comprises a variety of aspects (see figure 3) each containing aspect-specific contextual elements. Examples in this paper focus on the informational and the organizational aspect.

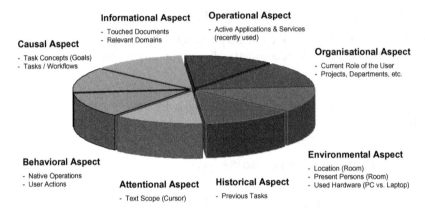

Fig. 3. In EPOS, the user's context comprises various aspects

We used Protégé[3] to realize an explicit and formal context model in RDF/S[4]. The classes, especially the `Context` class, and properties therein define the parameters for the user's context. An *actual* user context is represented by an instance of this context model, respectively, of the `Context` class. This instance, simply called *context object*, is permanently kept up-to-date. Hence, the *current* context of the user is then just a snapshot of the (one and only) context object and its contents. For every contextual information available, the context object contains so-called *contextual elements* wrapping the respective contextual information. For example: There could be some contextual elements wrapping the links to web pages the user recently browsed to while other contextual elements wrap some of the user's topics currently relevant.

Besides the core contextual information, these contextual elements also contain information about their confidence. Observed contextual information will have a confidence of 1.0 whereas estimated contextual elements, like automatically classified topics, will mostly have a lower confidence. Furthermore, contextual elements contain information about their creation and the support they give

---

[3] http://protege.stanford.edu/
[4] http://www.w3.org/TR/rdf-schema/

to or receive from other contextual elements. This allows for some explanation, as well as, for some global probability calculation on all of the elements in the user's context.

More important than the structure of the contextual element is the contextual information they convey. Before applications can use a context snapshot they have to *understand* the contextual information. As our context model is grounded in the user's personal knowledge space (see section 2.1), contextual elements represent real-life entities of the user's world. For example: touched documents, visited/classifying folders, recently touched or related persons, projects, and so on. This makes it easy for applications to understand and use the contextual information provided by our context service. Figure 7 shows a concrete real-life snapshot of the user's context. Only the informational and the organizational aspect have been extracted, e. g., behavioral information (actions of the user) has been elided to keep the RDF data as readable as possible.

It is important, that the context relies on the user's own personal knowledge space, i. e., his *individual and subjective* view of the world; therein for example the previously described topic ontology. This holds especially for the similarity of two user contexts. Their similarity is grounded on the similarity of the user's own view of the world. For example: For one user the topics workflow and business process modeling are the same while for a different user both topics are only similar. This means, the context model and contextual similarity adapt to the user.

4   Gathering and Eliciting Context

Context can be gathered and modeled using user observation and/or user feedback. However, relying on user feedback is critical. The context-sensitive support must be nearly perfect to keep even motivated users doing feedback all of the time. But still, even with perfect support, the user will stop giving feedback as soon as he is feeling stress and has no more time "to feed his context tamagotchi". Thus, *unobtrusive* and *reliable* gathering and modeling of context demands for permanent user observation. Of course the user must be able to toggle the user observation on and off whenever he likes.

Instead of integrating some low-level hooks into the operating system, EPOS created plugins for a set of standard applications from the scenario. The advantage is, that the plugins integrated in the applications can deliver more precise information than raw mouse movements, clicks, or key strokes. The price we pay is, that new applications can not be observed unless someone writes an observation plugin for it.

Again our scenario tells us, which applications we need to observe to gather the most interesting elements of the user context:

- File Explorer (handling documents and folders)
- Email Client (communication with colleagues)
- Web Browser (searching and browsing for documents/information)
- Text Processor (reading and writing documents)

User Observation for Mozilla Thunderbird (email client) and Mozilla Firefox (web browser) is realized in form of Mozilla extensions. That way they can be easily installed using an XPI-file. Besides, the user can configure and (de)activate the observation from within the application. File explorer and text processor plugins are still in development, but email communication and web browsing already allows for gathering very interesting contextual information. Anyway, all plugins send the observed user operations via simple XML-RPC calls to the listening *Context Service*, where the context object is updated accordingly. For example, when I opened an email in Mozilla Thunderbird, the following XML-RPC / HTTP POST is sent to `http://localhost:9998/RPC2`:

```
<methodCall>
  <methodName>epos_context_MozUserObsApi.eventViewEmail</methodName>
  <params><param><value><struct>
   <member><name>EmailURI</name>
     <value><string>imap://schwarz@serv-4100/INBOX#18423</string></value>
   </member><member><name>FolderURI</name>
    <value><string>imap://schwarz@serv-4100/INBOX</string></value>
   </member><member><name>Subject</name>
    <value><string>Personal and [...] - New Issue Alert</string></value>
   </member><member><name>Sender</name>
    <value><string>"alerts@springerlink.de</string></value>
   </member>
   ...
  </struct></value></param></params>
</methodCall>
```

Note, that the parameters are enclosed by `<![CDATA[...]]>`, really. For simplicity, these tags have been removed in the snippet above. After the XML-RPC has been sent, a servlet in the Context Server catches this event and calls the corresponding method:

```
public void eventViewEmail(Hashtable hashtable) {
  String emailURI = (String)hashtable.get("EmailURI");
  String subject = (String)hashtable.get("Subject");
  ...
  ViewEmail nop = new ViewEmail();  // nop stands for "native operation"
  nop.setURI(emailURI);
  nop.setSubject(subject);
  ...
  getContextServer.addNopToContext(nop);
}
```

We created an ontology of user operations using Protégé and RDF/S. The Java class `ViewEmail` used in the code snippet above has been generated by `rdf2java` [2] to wrap the corresponding RDF class in the Context Model's RDF Schema. All properties (outgoing edges) of some RDF object can be get and set by the Java wrapper. This allows very comfortable creation, handling, and passing of RDF data.

Besides parameterizing, the user operations are modeled hierarchically (is-a relations) to allow fine, as well as, coarse handling of user operations. For

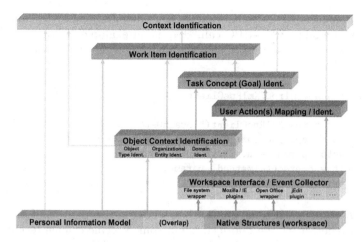

Fig. 4. The context elicitation is realized as a modular, pipelined architecture [13]

example: `ViewEmail` and `SendEmail` are `EmailOperations`; and `AddBookmark` is both a `BookmarkOperation` and an `ArchiveOperation`. A context-aware application can decide wether it is sufficient to know about the user doing some `EmailOperation` or wether `ViewEmail` and `SendEmail` leads to different user contexts and, thus, to different contextual support.

Adding the new *native operation* (*nop*) to the context means, first, creating a *ContextualElement* for the nop, and, second, raising a *ContextualElementAdded-Event*. This event will be catched by context elicitation modules. Taking the received user operations as evidences, more higher-level contextual information is estimated: Potentially relevant topics the user works on are inferred on the basis of recently touched documents (see section 2). Additionally, a case-based reasoning approach [13] estimates potential user goals and relevant workflow tasks. Figure 4 shows the context elicitation modules working in a pipelined architecture.

5   Retrieval and Usage of Context

In EPOS, context gathering and elicitation is realized as an autonomous, optional service. That service is used in two ways: *Context-Polling* and *Context-Listening*.

Context-Polling: Requesting a snapshot of the user's current context or a specific part of it. This is often triggered by some user function. The context can then be used, for instance, to show him recent or relevant items at that specific time. Another example: The user wants to save a downloaded document. The current context can be used to propose a corresponding directory. Furthermore, if the document is annotated with the context, the document can be retrieved via a context-enabled query later. From inside the EPOS framework, the context model can be retrieved simply by calling

```
Model model = ContextService.getContextApi().getSnapshot();
```

From outside the EPOS framework, or from non-Java code, the same can be achieved by a simple XML-RPC call. Here is an example of how to request a context snapshot from some Java application – the result will an output similar to that in figure 7:

```
import org.apache.xmlrpc.XmlRpcClient;
...
XmlRpcClient c = new XmlRpcClient("localhost", 9998);
String method = "epos_context_ContextApi.getSnapshotAsRdfXml";
java.util.Vector params = new java.util.Vector();
Object result = c.execute(method, params);
System.out.println("context as RDF/XML:\n" + result);
```

Context-Listening: If some application/service registers at the Context Service as a listener, it will be informed about some specific change (event) in the user's context. On one hand, this is already heavily used internally by the context elicitation modules. They are triggered as soon as new contextual elements enter the context. On the other hand, this is needed to realize *pro-active* support. If the user shifts his actions towards some context for which relevant support is available, this support can be presented to the user without him having to asking for it. The context listening can also be used to trigger some context-sensitive help system, for instance, a personal (knowledge) work assistant. By now, the context listening protocol is quite proprietary and can only be used by modules *inside* the EPOS framework, but extending event listening to XML-RPC clients is just a minor, technical issue.

6   Applications of Context

This section presents some concrete applications created in the EPOS scenario. All these applications handle context in a different way, however, they all access the user's context via the interface described in section 5.

6.1   EPOS Assistant Bar

The so-called *EPOS Assistant Bar* is a graphical user interface (GUI) delivering potentially relevant information items to the user in a pro-active way. Despite the envisioned pro-active support, the user should be able to work on as undisturbed as possible. Therefore, the design of the GUI has to face three challenges: First of all, it must not show irrelevant information to the user as this would only lead to distraction and annoyance. Second, the user must be able to classify, evaluate, and understand the presented information at first sight. And third, the presentation itself should omit any form of distraction, e.g., the mere presentation of information should be sufficient. Usage of animation, blinking, or even sound must not be used.

The assistant bar is always present above the Windows task bar. It never changes its size or visibility. It never tries to attract the user's intention besides

Fig. 5. The EPOS Assistant Bar is positioned directly above the Windows task bar. Its four panes show currently relevant information items. From left to right these panes show: documents, categories, organizational entities, and tasks.

the mere display of (new) information. The geometrical positioning of the presented information together with the usage of small icons enables a fast categorization for the user. The assistant bar uses the following four panes to distinguish context-relevant information graphically: The left-most pane shows documents (or resources with document character); directly to the right there are categories (topics); the next pane shows organizational entities (projects, persons, contacts); and the right-most pane proposes task-oriented items (workflow tasks and events). See figure 5 for a screenshot of the EPOS Assistant Bar.

6.2 EPOS Notes

A small application called *EPOS Notes* allows quick and easy note-taking. The most important point here is, that entered notes are annotated with the user's context. The accessibility of a note's *creation context* enables (a) context-sensitive recherche and (b) context-triggered presentation of notes. For example, a note entered in the context of browsing a concrete web page can be displayed if the user browses the same web page again. The presentation of the note is done using the EPOS Assistant Bar (see section 6.1).

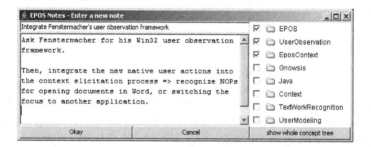

Fig. 6. The EPOS Notes provides semi-automatic classification of small text notes. Furthermore, the entered note is annotated with the user's current context.

Furthermore EPOS Notes shares its data structures with a web log (we use WordPress[5]). This means first of all, that entering and searching for notes can also be done using the web log's frontend, i. e., the user enters and searches his web log entries. But that's not all. Melting notes and blog further means, that notes entered with EPOS Notes are also web log entries. There is no conversion

---

[5] http://wordpress.com/

needed. A new note is directly entered as a web log entry. The only difference is (currently) that, only these notes entered with EPOS Notes receive the additional creation context, but that is just a technical issue.

Actually, there is another difference to a standard web log: EPOS Notes applies a document classification approach to automatically and continually classify an entered note while the user types it. The resulting categories for the note are proposed to the user who can decide which of these are correct and which are missing (see figure 6).

6.3   Context-Sensitive Bookmark Service/Portal

The EPOS user observation framework includes the detection and storage of the user's browsing behavior. The *resource statistics* component of EPOS uses this to store document usage statistics like the *amount* of viewing operations (measured in number of days), the *frequency* (average number of days between two accesses), and *neighborhood* (documents used in the same context, i. e., shortly before or after each other).

Additionally, another EPOS component, the so called *bookmark-service*, annotates each viewed web page with the user's current context. For instance, if the user is currently working for some project. The viewed web page will be related to that project, too. Analogously to EPOS Notes this allows context-sensitive retrieval and filtering of viewed web pages. One small conceptual difference to EPOS Notes is, however, that the context annotation takes place automatically for every *visited* web page and not only for some selected ("bookmarked") pages. That way, web pages viewed, for example, during a recherche process will not be lost but can be retrieved using time- or context-specific queries. A graphical user interface providing such a context-sensitive bookmark portal is currently being developed.

6.4   Context-Sensitive Classification Service

Besides the classification and retrieval of notes or bookmarks. There are a lot of use cases which benefit from an automatic elicitation of the user's context. In particular, the user context contains information about categories potentially relevant for the user. These can be used to classify new, emerging information items. For example, a downloaded PDF document can be classified using the currently relevant categories in the user context.

EPOS realized a so called *Drop Box*, which is a file folder together with a service listening on changes of the folder's contents. Whenever the user drops a file there, e. g., by downloading something there, the Drop Box proposes the categories currently in the user context.[6] Some of these proposed categories may represent file folders the user created earlier. In that case, the drop box offers the *"Move and Classify"* functionality. The user can choose *one* category

---

[6] If the Drop Box manages to retrieve the *text* content of some dropped document, it calculates relevant categories for the text content using a document classification tool and proposes these together with the categories found in the user context.

representing a file folder where the file is *moved* to. In parallel, he can choose *several* other categories which will used to *classify* the file. Here, *moving* a file means actually moving the file in the file system while *classifying* a file is delegated to our document classification tool (BrainFiler). The classification of the file re-trains the internal feature vector of the category due to the just added file. So, each classify action boosts the classification itself.

6.5   Context-Aware Workflow Interface

The user context model includes enough compatible contextual elements to enable a context-aware workflow interface. One component of this interface will identify workflows/tasks relevant for the user's current context [14] and propose relevant tasks together with relevant information annotated to these tasks. That way the valuable information stored in a workflow management system (WfMS) will be (re)used to assist a worker without him having to trigger the WfMS explicitly. Besides the pro-active information assistance, a context-sensitive view (filter) on the work list can reduce the number of tasks to an until the worker is able to overview it. The context-relevant retrieval of tasks can also be extended to tasks of colleagues to encourage more collaboration among follow workers.

The context-aware workflow interface is currently under construction. The pro-active presentation of context-relevant material stored in the WfMS will be integrated into the EPOS Assistant Bar as described in section 6.1.

7   Related Work

The Lumiere project [9] realized pro-active assistance for computer software users. A bayesian user model was employed to infer a user's information needs. While Lumiere mainly focussed on enhancing the usage of some software, EPOS pursues assisting the users' work in general. That means, first, not only one application is observed and enhanced with some (closed) assistance, and second, not only the user's *information need* is elicited, but his context.

Watson [3] observes user interaction with everyday applications and attempts to automatically fulfil a user's information need by querying common Internet information sources automatically. Analogously to Watson, EPOS uses a *Content Analyzer* to automatically classify some viewed document and estimate contextually relevant topics that way. Furthermore, EPOS also realizes a real task-specific information delivery using the information stored in a workflow management system. It knows about which documents are relevant for which task. As the context also contains relevant workflow tasks, the corresponding documents can be presented to the user [8].

The remembering agent *Margin Notes* [10] proposes relevant documents matching the user's browsing behavior. The documents are presented in the web browser, particularly, in an additional side bar right to the displayed web page. The interesting thing here is, that Margin Notes puts information of relevant material directly adjacent to the relevant paragraph of the web page. Besides the restriction of the user observation and context elicitation to the web scenario, Margin Notes embodies a useful, pro-active and quite unobtrusive assistance interface which is similar to the EPOS Assistant Bar (see section 6.1). However, our assistant bar lacks the nice feature of graphically showing relevancy correlations of text passages and relevant material.

Fenstermacher [7] envisions revealing and storing process-relevant information by automatically classifying documents the user touches during his work. For this purpose low-level observation hooks are installed directly into the operating system. EPOS relies on observed user actions with higher semantics, like `SendEmail`, however, collaborating with Fenstermacher we may integrate these low-level hooks to be able to observe applications without observation plugins (see section 4).

Zacarias et al. [15] introduce an operating systems metaphor, where the user's *operating system* acts as an *engine* managing the execution of pre-defined flows of work. Besides that very interesting metaphor, the authors carry out a case study where students observe and record the actual tasks, actions, and interactions of persons at work. EPOS will keep up collaboration with Zacarias' team to enhance the context model according to their observations.

8   Conclusion

We have presented an explicit and formal model of a knowledge worker's context. The user's context, being an instance of this context model, is fed by user observation plugins installed in standard applications like Mozilla Thunderbird and Mozilla Firefox.

The context relies on the user's personal workspace, i. e., touched or elicited resources in the context are mainly the user's own files, folders, email contacts, and so on. This means, that the contextual elements are known entities and, hence, the context can be easily understood and used. It also means, that the elicited context adapts to the specific user at a specific time. Even if the user changes his working procedures and domain over time, the context will adapt automatically, because according changes on the user's PC will take place: new folders and new documents emerge, re-classification of documents occurs, etc.

An application can act context-aware by either explicitly requesting a context snapshot or by registering itself as a listener to context changes. Several context-aware applications have been presented in section 6. In particular, we envision the realization of a pro-active but unobtrusive knowledge assistant. The goal is, to support the individual worker by proposing material he knows and/or understands, that is, documents, categories or structures he or his colleagues created during work. This allows him to get his job done quicker and without losing his flow that often.

On the other hand, such assistance is only meaningful if the available material and structures are good and up-to-date. Some of the structures, especially file folders, are already present at a user's desktop while others have to be created separately, be it manually or (semi-)automatically. Typical top-down knowledge management (KM) approaches oblige the worker to enter the missing data in

```
<rdf:RDF
    xmlns:rdf="http://www.w3.org/1999/02/22-rdf-syntax-ns#"
    xmlns:rdfs="http://www.w3.org/2000/01/rdf-schema#"
    xmlns:context="http://km.dfki.de/context#"
    xmlns:domain="http://km.dfki.de/domain#"
    xmlns:org="http://km.dfki.de/org#"
    xmlns:object="http://km.dfki.de/nasti/objects#">
  <context:Context>
    <context:informationalAspect>
      <context:InformationalAspect>
        <context:contains>
          <object:EMail rdf:about="imap://schwarz@serv-4100/INBOX/;UID=17050">
            <object:subject>paper on data integration framework, ISWC 2005</object:subject>
            <object:recipients>Sven Schwarz &lt;schwarz@dfki.uni-kl.de&gt;</object:recipients>
            <object:sender>Leo Sauermann &lt;leo@gnowsis.com&gt;</object:sender>
            <object:content>Hi Sven! [...] I would like to write a paper about [...]
                            semantic web and desktop applications [...]
            </object:content>
            <object:lastAccess>2005-04-13T14:45:22</object:lastAccess>
            <context:confidence>1.0</context:confidence>
          </object:EMail>
        </context:contains>
        <context:contains>
          <object:HtmlFile rdf:about="http://jena.sourceforge.net/">
            <object:location>http://jena.sourceforge.net/</object:location>
            <object:title>Jena Semantic Web Framework</object:title>
            <object:fileType>text/html</object:fileType>
            <object:lastAccess>2005-04-13T14:45:42</object:lastAccess>
            <context:confidence>1.0</context:confidence>
          </object:HtmlFile>
        </context:contains>
        <context:contains>
          <domain:DomainConcept rdf:about="urn:brainfiler:dfkiklkm:schwarz:Category:136">
            <domain:name>Gnowsis</domain:name>
            <context:confidence>0.717095</context:confidence>
            <context:supportedBy rdf:resource="http://jena.sourceforge.net/"/>
            <context:supportedBy rdf:resource="imap://schwarz@serv-4100/INBOX/;UID=17050"/>
          </domain:DomainConcept>
        </context:contains>
        <context:contains>
          <domain:DomainConcept rdf:about="urn:brainfiler:dfkiklkm:schwarz:Category:105">
            <domain:name>RDF[S]</domain:name>
            <context:confidence>0.350096</context:confidence>
            <context:supportedBy rdf:resource="http://jena.sourceforge.net/"/>
          </domain:DomainConcept>
        </context:contains>
      </context:InformationalAspect>
    </context:informationalAspect>
    <context:organizationalAspect>
      <context:OrganizationalAspect>
        <context:contains>
          <org:Person rdf:about="mailto:leo@gnowsis.com">
            <org:eMail>leo@gnowsis.com</org:eMail>
            <org:firstName>Leo</org:firstName>
            <org:lastName>Sauermann</org:lastName>
            <context:confidence>1.0</context:confidence>
            <context:supportedBy rdf:resource="imap://schwarz@serv-4100/INBOX/;UID=17050"/>
          </org:Person>
        </context:contains>
      </context:OrganizationalAspect>
    </context:organizationalAspect>
  </context:Context>
</rdf:RDF>
```

Fig. 7. Snapshot of the user's context. Only the informational and the organizational aspect have been extracted.

order to get the KM tools running. However, we are aiming at eliciting as much as possible automatically. For areas where this is not possible, our bait is to compensate the worker for his additional work. For example, the user will model a new project and its project members in the *organizational repository* freely if the context elicitation is then able to proposes the new project together with related information like the project homepage for example.

The set of supported applications and, accordingly, the set of observable user actions is limited by the set of available user observation plugins. Implementing new plugins for yet unsupported applications is merely a question of how to integrate plugins into that application and how to recognize the the user's actions therein. We plan to integrate native user observation modules into the context elicitation framework in order to get a more complete and accurate picture of the user's context. For example, recognizing an application switch provides evidences a context switch of the user.

The currently implemented context elicitation modules elicit *projects*, *persons*, and *topics* potentially relevant due to the text content created or viewed by the user. Ongoing research will cover the elicitation of higher-level contextual information such as the user's *goals* or relevant *workflow tasks*. The identification of relevant workflow tasks enables a pro-active information delivery provided indirectly by a workflow management system (WfMS). That way, information stored in a WfMS will be (re)used more often than in a classical workflow management environment.

The ongoing research concerning the context elicitation methodology, as well as, the implementation of context-aware assistance applications will be followed by an evaluation of the fitting and utility of the elicited context. Furthermore, the evaluation will have to show whether and how much the users really like and use the context-aware support.

References

1. Homepage of FRODO TaskMan: http://www.dfki.de/frodo/taskman/.
2. Homepage of rdf2java: http://rdf2java.opendfki.de/.
3. Jay Budzik and Kristian J. Hammond. Watson: An infrastructure for providing task-relevant, just-in-time information.
4. Andreas Dengel, Andreas Abecker, Jan-Thies Bähr, Ansgar Bernardi, Peter Dannenmann, Ludger van Elst, Stefan Klink, Heiko Maus, Sven Schwarz, and Michael Sintek. Evolving Personal to Organizational Knowledge Spaces. Project Proposal, DFKI GmbH Kaiserslautern, 2002.
5. Ludger van Elst and Andreas Abecker. Integrating Task, Role, and User Modeling in Organizational Memories. In *14 Int. FLAIRS Conference, Special Track on Knowledge Management, Key West, Florida, USA*, May 2001.
6. Ludger van Elst, Felix-Robinson Aschoff, Ansgar Bernardi, Heiko Maus, and Sven Schwarz. Weakly-structured workflows for knowledge-intensive tasks: An experimental evaluation. In *IEEE WETICE Workshop on Knowledge Management for Distributed Agile Processes: Models, Techniques, and Infrastructure (KMDAP03)*. IEEE Computer Press, 2003.
7. Kurt D. Fenstermacher. Revealed Processes in Knowledge Management. In Klaus-Dieter Althoff, Andreas Dengel, Ralph Bergmann, Markus Nick, and Thomas Roth-Berghofer, editors, *3rd Conference on Professional Knowledge Management – WM 2005*, pages 397–400. DFKI GmbH, 2005.
8. Harald Holz and Frank Maurer. Knowledge management support for distributed agile software processes. In *Advances in Learning Software Organizations, 4th International Workshop, LSO 2002, Chicago, IL, USA, August 6, 2002, Revised Papers.*, volume 2640. Springer, 2002.

9. E. Horvitz, J. Breese, D. Heckerman, D. Hovel, and K. Rommelse. The lumiere project: Bayesian user modeling for inferring the goals and needs of software users. In *In Proceedings of the Fourteenth Conference on Uncertainty in Artificial Intelligence*, pages 256–265, Madison, WI, July 1998.

10. Bradley J. Rhodes. Margin notes: Building a contextually aware associative memory. *The Proceedings of the Internatial Conference on Intelligent User Interfaces(IUI '00)*, 2000.

11. Leo Sauermann. The gnowsis-using semantic web technologies to build a semantic desktop. Diploma thesis, Technical University of Vienna, 2003.

12. Leo Sauermann and Sven Schwarz. Introducing the gnowsis semantic desktop. In *Proceedings of the International Semantic Web Conference 2004*, 2004.

13. Sven Schwarz and Thomas Roth-Berghofer. Towards goal elicitation by user observation. In *Proceedings of the FGWM 2003 Workshop on Knowledge and Experience Management*, Karlsruhe, 2003.

14. Roza Shkundina and Sven Schwarz. A similarity measure for task contexts. In *Proceedings of the Workshop Similarities - Processes - Workflows in conjunction with the 6th International Conference on Case-Based Reasoning*, Chicago, 2005.

15. Marielba Zacarias, Artur Caetano, H. Sofia Pinto, and Jose Tribolet. Modeling contexts for business process oriented knowledge support. In *Proceedings of the WM 2005 Workshop on Knowledge Management for Distributed Agile Processes (KMDAP 2005)*, Kaiserslautern, 2005.

Situation Modeling and Smart Context Retrieval with Semantic Web Technology and Conflict Resolution

Dominik Heckmann

German Research Center for Artificial Intelligence
heckmann@dfki.de

Abstract. We present a service to model situations and retrieve contextual information in mobile and ubiquitous computing environments. We introduce the general user model and context ontology GUMO for the uniform interpretation of distributed situational information in intelligent semantic web enriched environments. Furthermore, we present the relation to the user model and context markup language USERML, that is used to exchange partial models between different adaptive applications. Our modeling and retrieval approach bases on semantic web technology and complex conflict resolution concepts.

Keywords: situation modeling, smart context retrieval, user model and context ontology, semantic web, distributed situation service, user model markup language.

1 Motivation and Introduction

Increasing mobility of interactive systems like portable and wearable computers and ubiquitous computing with embedded intelligent devices in everyday objects make context-aware, situation-aware and especially user-adaptive computing more important. Systems that adapt to their users or contexts need to have access to information about them. However, most currently implemented human computer interaction systems that employ a user model or context model work with isolated models.

The main challenge in our opinion is to let different systems that are interconnected or even organized in intelligent environments, communicate about their context models and user models. This challenge has been motivated by the expected result that *permanent evaluation of user behavior with different systems and devices will lead to better models and thus allow better functions of adaptation like adaptive web-sites, recommended products, adaptive route planning or better speech interaction.*

This motivation points towards the interrelation of the research areas *user-adaptivity*, *context-awareness*, *ubiquitous computing* and *semantic web*, whith the goal to use semantic web technology as mediator between the other three areas. The title "Situation Modeling and Smart Context Retrieval" seems to be inconsistent. However, we divide between *situation* modeling and *context* retrieval on purpose because at modeling time, we can not know which of the situational information will be considered as being context. Thus, context in this sense is retrieval dependent.

This paper consists of five parts. The first one presents our underlying model of situated interaction with context-awareness and user-adaptivity. The second one

T.R. Roth-Berghofer, S. Schulz, and D.B. Leake (Eds.): MRC 2005, LNAI 3946, pp. 34–47, 2006.

presents a conceptual overview of the overall service architecture that has been realized as a web service like information broker. The third part briefly describes our approach to *Situation Modeling* with the markup language UserML and the user model and context ontology GUMO. The fourth part introduces the *Smart Context Retrieval* that bases on information retrieval with integrated semantic conflict resolution. Applications and interfaces that use and control the context information are presented in the last part. The whole approach has been implemented and can be tested at http://www.u2m.org/.

2   Integrated Model for Context-Awareness and User-Adaptivity

Situated interaction takes places between a user and a system in a surrounding environment. In [1] it is pointed out that throughout the different research communities and disciplines, there are various definitions of what exactly is contained in the *context model* [2], the *user model* [3], and the *situation model* [4]. Therefore, it is necessary to clarify how those terms will be used in our approach. A *situation model* is defined in our approach as the combination of a *user model*, a *context model* and a *resource model*. Figure 1 presents a diagrammatic answer to the question: *What is situated interaction and how can we conceptualize it?*

In our approach, all information that is stored in this situation model is represented in the datastructure of so called SITUATIONALSTATEMENTS, see section 4 and [5], that collect apart from the main descriptive information also meta data like temporal and spatial constraints, explanation components and privacy preferences. Distributed sets of so called SITUATIONREPORTS form a coherent, integrated, but still hybrid accretion

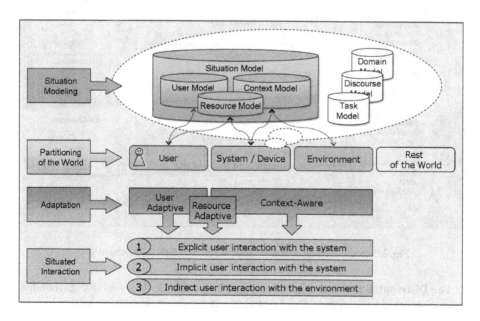

Fig. 1. Situated interaction and the system's situation model for mobile computing

concept of ubiquitous situation models. Before we look at this modeling point, we first present the overall service architecture.

3   Architecture of a Highly Distributed SITUATIONSERVICE

Our so called u2m.org SITUATIONSERVICE manages information about users, about their contexts and the situation in general[1]. It contributes additional benefit compared to a pure situation *server* or context *broker* that only manage information. The presented service is an independent application with a distributed approach for accessing and storing information, the possibility to exchange and understand data between different applications as well as adding privacy and transparency to statements. A key feature is that the semantics for all user model and context dimensions are mapped to the general user model & context ontology GUMO, see section 4 or [6]. Thus, the inter-operability between distributed user-adaptive and context-aware systems is granted. Figure 2 shows the main actors and modules of the u2m.org SITUATIONSERVICE.

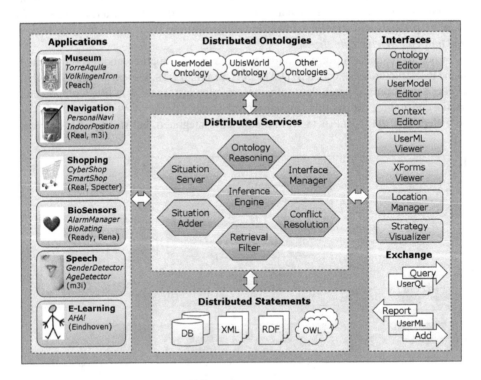

Fig. 2. Modularized architecture of the u2m.org SITUATIONSERVICE

The **Distributed Services** box is presented in the middle. It is literally surrounded by its connected environment. Even though the items are shown conceptually close to

---

[1] Our SITUATIONSERVICE can be tested at http://www.u2m.org

each other, they are spatially spread throughout the whole scenery. The box contains a set of modules that represent tasks and roles that are offered by the service:

- *Situation Server*, a web-server that manages the storage of the statements
- *Situation Adder*, a parser that analyzes incoming statements and distributes the repositories.
- *Retrieval Filter*, a procedure that controls the retrieval of situation statements
- *Conflict Resolution*, a complex process that detects and resolves possible conflicts
- *Inference Engine*, a proactive engine that applies meta rules and triggers events
- *Interface Manager*, a control mechanism that integrates the user interfaces
- *Ontology Reasoning*, a reasoner that applies knowledge from the various ontologies

The **Applications** box on the left, sorts the applications that already cooperate with the SITUATIONSERVICE according their application domain: *museum, navigation, shopping, biosensors, speech* and *e-learning*.

The **Distributed Statements** box on the bottom points to the clear separation between data and software. The repositories are completely independent from the services which allows various services to operate independently on the same knowledge bases. This is only possible because the privacy and administration attributes are attached to each SITUATIONALSTATEMENT and not (as in most other systems) handled by the broker system.

The **Distributed Ontologies** box on the top points to the clear separation between the syntax and the semantics as discussed in the following section. These ontologies are used for the interpretation of statements, for the detection of conflicts and for the definition of expiry defaults and privacy defaults.

The **Interfaces & Exchange** box on the right points to the clear separation between the service and the user interfaces and development tools which results in the advantage that each interface and tool can operate with different repositories, different ontologies and even different services. This is for example important for the spatially spread computational setting within ubiquitous computing. If the network connection is lost, the user interfaces can smoothly swap to device-local systems or integrate spatially restricted repositories. The communication between the boxes and items is indicated by the bipolar arrows. UserQL is used to ask the queries, UserML is used to report the answers and to add new statements.

Figure 3 shows the input and output information flows *add*, *request* and *report* of the SITUATIONSERVICE. They are denoted as (yellow) arrows. The numbers in the (orange) ovals present the procedural order. Number (1) visualizes the sensors, users and systems that add statements via UserML. The statements are sent to the so called *Situation Adder*, a parser that preprocess the incoming data and distributes them to the different repositories, as indicated by number (2). If now a request is sent to the *Situation Server* via UserQL from a user or a system, see number (3), the repositories are selected from which the statements are retrieved as shown at number (4.1). Then conflict resolution strategies are applied, see number (4.2), and the semantic interpretation as indicated by number (4.3). Finally, see number (5), the adapted output is formatted and sent via HTTP in form of an UserML report back to the requesting user or system.

Even though, the description of the distributed SITUATIONSERVICE so far is very brief it should be understood as basic framework for the exchange and storage of sit-

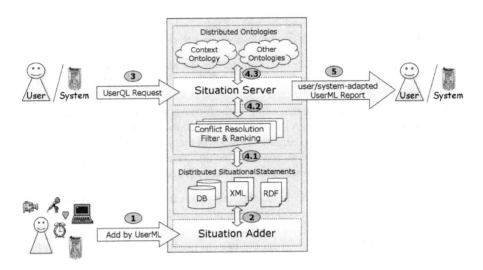

Fig. 3. General procedural view to the SITUATIONSERVICE

uational information, the knowledge representation techniques that are defined in the following section.

4   Situation Modeling with UserML and GUMO

Ontologies provide a shared and common understanding of a domain that can be communicated between people and heterogeneous and widely spread application systems, as pointed out in [7]. Since ontologies have been developed and investigated in artificial intelligence to facilitate knowledge sharing and reuse, they should form the central point of interest for the task of exchanging user models. The user model & context markup language USERML is defined as an XML application, see [8]. However, XML is purely syntactic and structural in nature. The RDF standard has been proposed as a data model for representing meta data by [9]. Nonetheless, the web ontology language OWL has more facilities for expressing semantics, [10]. OWL can be used to explicitly represent the meaning of terms in vocabularies and the relationships between those terms. Thus, OWL is our choice for the representation of user model and context dimension terms and their interrelationships. This ontology should be available for all user-adaptive and context-aware systems at the same time, which is perfectly possible via internet and wireless technology. The major advantage would be the simplification for exchanging information between different systems. The current problem of syntactical and structural differences between existing adaptive systems could be overcome with such a commonly accepted ontology. GUMO[2] is collecting the user's dimensions that are modeled within user-adaptive systems like the *user's heart beat*, the *user's age*, the *user's current position*, the *user's birthplace* or the *user's ability to swim*. Furthermore, the modeling of the user's interests and

---

[2] GUMO homepage: http://www.gumo.org

preferences like *reading poems* or *playing adventure games* is analyzed. Secondly, the contextual dimensions like *noise level* in the environment, *battery status* of the mobile device, or the outside *weather* conditions are modeled.

The main conceptual idea in SITUATIONALSTATEMENTS, is the division of the user model and context dimensions into the three parts: `auxiliary`, `predicate` and `range` as shown below.

$$\texttt{subject} \left\{ \textit{UserModel\&ContextDimension} \right\} \texttt{object}$$
$$\Downarrow$$
$$\texttt{subject} \left\{ \texttt{auxiliary, predicate, range} \right\} \texttt{object}$$

Fig. 4. From the RDF triples to five tuples

Apart from these five so called `mainpart` attributes, there are predefined attributes about the `situation`, the `explanation`, the `privacy` and the `admini-`

Situational Statement / XML (Mix)	Situational Statement / Box	
`<statement>`	**Mainpart**	
`    <mainpart`	Subject	= Flight LH225
`        subject = "a1"`	Auxiliary	= hasPlan
`        auxiliary ="a2"`	Predicate	= boarding
`        predicate = "a3"`	Range	= time
`        range = "a4"`	Object	= in 10 minutes
`        object = "a5" />`	**Situation**	
`    <situation`	Start	= 2003-04-16T19:14
`        start = "a6"`	End	= ?
`        end = "a7"`	Durability	= few minutes
`        durability = "a8"`	Location	= airport.gate23
`        location = "a9"`	Position	= ?
`        position = "a10" />`	**Explanation**	
`    <explanation`	Source	= airport.repository
`        source = "a11"`	Creator	= airport.inference
`        creator = "a12"`	Method	= deduction13
`        method = "a13"`	Evidence	= fight-system
`        evidence = "a14"`	Confidence	= 0.6
`        confidence = "a15" />`	**Privacy**	
`    <privacy`	Key	= ********
`        key = "a16"`	Owner	= Airport
`        owner = "a17"`	Access	= public
`        access = "a18"`	Purpose	= commercial
`        purpose = "a19"`	Retention	= 1 month
`        retention = "a20" />`	**Administration**	
`    <administration`	id	= 3
`        id = "a21"`	unique	= u2m.org#154125
`        unique = "a22"`	replaces	= u2m.org#152148
`        replaces = "a23"`	group	= ContextModel
`        group = "a24"`	notes	= ;-)
`        notes = "a25" />`		
`</statement>`		
(1)	(2)	

Fig. 5. (1) A SITUATIONALSTATEMENT defined as XML document with the five parts: mainpart, situation, explanation, privacy and administration. (2) An example: contextual information for travelers at an airport: the boarding for the flight is in 10 minutes.

stration as shown in figure 5 which presents the concept of SITUATIONAL-STATEMENTS in an XML instantiation and which gives a first example of such statements.

Our basic context models are more expressive than simple attribute-value pairs or RDF triples. If one wants to say *something about the user's interest in football*, one could divide this into the `auxiliary`=*hasInterest*, the `predicate`=*football* and the `range`=*low-medium-high*. If one wants to express something like *knowledge about symphonies*, one could divide this into the `auxiliary`=*hasKnowledge*, the `predicate`=*symphonies* and the `range`=*poor-average-good-excellent*.

GUMO is designed according to this USERML approach. Approximately one thousand groups of `auxiliaries`, `predicates` and `ranges` have so far been identified and inserted into the ontology. However, it turned out that actually everything can be a `predicate` for the `auxiliary` *hasInterest* or *hasKnowledge*, what leads to a problem. The suggested solution is to identify basic user model dimensions on the one hand while leaving the more general world knowledge open for already existing other ontologies on the other hand. Candidates are the general suggested upper merged ontology SUMO, see [11], and the UBISONTOLOGY[3], see [12], to model intelligent environments. Which groups of user dimensions can be identified? In [13] and [14] rough classifications for such categories can be found. Identified user model and context `auxiliaries` are for example *hasKnowledge, hasInterest, hasBelieve, hasPlan, hasProperty, hasGoal, hasPlan, hasRegularity* and *hasLocation*. We restrict ourself in this paper to present user model `predicates` that fit to the `auxiliary`: *hasProperty*, the so called *BasicUserDimensions*.

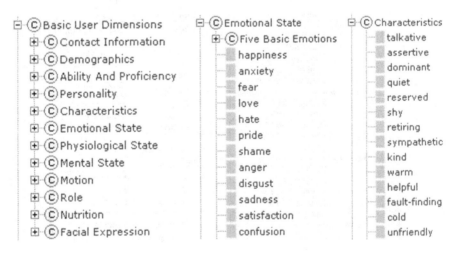

Fig. 6. Some *BasicUserDimensions*: Emotional States, Characteristics and Personality. The complete ontology can be inspected with a foldable tree browser at www.gumo.org

The listing in figure 7 presents the concept *PhysiologicalState* defined as `owl:Class`. It is defined as a subclass of *BasicUserDimensions*. A class defines a

---

[3] UbisWorld homepage: http://www.ubisworld.org

```
<owl:Class rdf:ID="PhysiologicalState.700016">
  <rdfs:label> Physiological State </rdfs:label>
  <rdfs:subClassOf rdf:resource="#BasicUserDimensions.700002" />
  <gumo:identifier> 700016 </gumo:identifier>
  <gumo:lexicon>state of body or bodily functions</gumo:lexicon>
  <gumo:privacy> high.640033 </gumo:privacy>
  <gumo:website rdf:resource="&GUMO;concept=700016" />
</owl:Class>
```

Fig. 7. The OWL class definition of "Physiological State"

group of individuals that belong together because they share some properties. Classes can be organized in a specialization hierarchy using `rdfs:subClassOf`.

Every concept has a unique `rdf:ID`, that can be resolved into a complete URI. Since the handling of these URIs could become very unhandy, a short identification number was introduced, the so called `gumo:identifier`. The identification number has the advantage of freeing the textual part in the `rdf:ID` from the need of being syntactically unique. Apart from solving the problem of conceptual ambiguity, this number facilitates the work within relational databases, which is important for the implementation. The lexical entry `gumo:lexicon` is defined as *the state of the body or bodily functions*, while it could also be realized through a link to an external lexicon like WORDNET. The attribute `gumo:privacy` defines the default privacy status for this class of user dimensions. It can be overridden in the concrete SITUATIONALSTATEMENT. The attribute `gumo:website` points towards a web site, that has its purpose in presenting this ontology concept, to a human reader. The abbreviation `&GUMO;` is a shortcut for the complete URL to the GUMO ontology in the semantic web.

The listing in figure 8 defines the dimension *Happiness* as an `rdf:Description`. The attribute `gumo:expiry` provides a default value for the average expiry which carries the qualitative time span of how long the statement is expected to be valid. In most cases when user model and context dimensions are measured, one has a rough idea about the expected expiry. For instance, emotional states hold normally no longer than 15 minutes, however personality traits won't change within months. Since this qualitative time span is dependent from every user model dimension, it should be defined within GUMO.

Another important point that is shown here is the ability of multiple-inheritance in OWL. In detail, *happiness* is defined as `rdf:type` of the class *EmotionalState* and *FiveBasicEmotions*. Thus OWL allows to construct complex, graph-like hierarchies of

```
<rdf:Description rdf:ID="Happiness.800616">
  <rdfs:label> Happiness </rdfs:label>
  <rdf:type rdf:resource="#EmotionalState.700014" />
  <rdf:type rdf:resource="#FiveBasicEmotions.700015" />
  <gumo:expiry> 15 minutes </gumo:expiry>
  <gumo:image rdf:resource="http://u2m.org/img/happiness.gif" />
</rdf:Description>
```

Fig. 8. The OWL definition of "Happiness"

user model concepts, which is especially important for ontology integration. Some examples of rough expiry-classifications are:

- physiologicalState.heartbeat - can change within seconds
- mentalState.timePressure - can change within minutes
- characteristics.inventive - can change within months
- personality.introvert - can change within years
- demographics.birthplace - can't normally change at all

The idea behind gumo:expiry is that if no new value is available on the SITUATION-SERVICE one can still work with old values, combined with reduced confidence values.

To support the distributed construction and refinement of GUMO, we developed a specialized online editor to introduce new concepts, to add their definitions and to transform the information automatically into the required semantic web language. The GUI of this online editor can be be found at www.ubisworld.org. This editor is part of a whole set of online tools that are collected under the name *UbisWorld*. An example screenshot can be found in figure 10. The left-hand side offers an hierarchical dropdown list and the right-hand side offers a detailed page for one instance of this world.

5   Smart Context Retrieval with Semantic Conflict Resolution

The architectural diagram in figure 9 shows the SMARTSITUATIONRETRIEVAL or smart context retrieval process. The focus is set on the semantic conflict resolution part.

The oval numbers indicate the reading direction. Item (1) shows the request in UserQL that has to be parsed first. Item (2) points to the distributed retrieval of SITUATIONALSTATEMENTS. Item (3) summarizes the three macro-steps select, match and filter and presents the FILTERINGRESULT as input to the conflict resolution process. Item (4) stands for the three syntactical procedures VARIATIONMAPPING, REMOVEEXPIRED and REMOVEREPLACED. Item (5) shows the

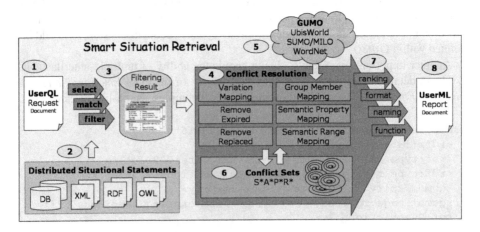

Fig. 9. Smart Context Retrieval with Focus on Semantic Conflict Resolution

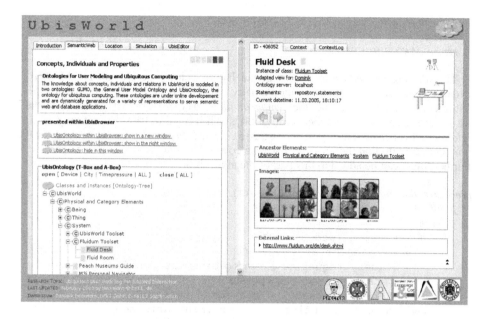

Fig. 10. UbisWorld Fluid Desk

three semantical procedures GROUPMEMBERMAPPING, SEMANTICPROPERTYMAP-
PING and SEMANTICRANGEMAPPING that base on knowledge in the ontologies
GUMO, UbisWorldOntology, SUMO/MILO and the knowledge base WorldNet. Item
(6) shows the detection of syntactic and semantic conflicts and the construction of
$\langle S^*, A^*, P^*, R^* \rangle_{nonReplaced}^{nonExpired}$ conflict sets. Item (7) points to the post-processing of
ranking, format, naming and function that control the output format. Item
(8) forms the resulting UserML report, that is sent via HTTP to the requestor.

The *matching procedure* as shown in the listing below compares all given match attributes with the corresponding statement attributes. Furthermore it integrates semantic
functionality, here shown by ontological *extension* and spatial *inclusion*.

procedure matching (query, SELECTEDREPOSITORY)
   forall statement ∈ REPOSITORY
     begin
       if (statement.subject = query.subject
         or statement.subject ∈ *extension*(query.subject))
         ∧ statement.auxiliary = query.auxiliary
         ∧ (statement.predicate = query.predicate
         or statement.predicate ∈ *extension*(query.predicate))
         ∧ statement.range = query.range
         ∧ statement.object = query.object
         ∧ statement.id = query.id
         ∧ statement.group = query.group
         ∧ (statement.location = query.location
         or statement.location ∈ *inclusion*(query.location))
         ∧ ...

>   **then**   **add** statement **to** MATCHINGRESULT
>   **end**
> **return** MATCHINGRESULT

The *filtering procedure* as shown in the listing below operates on the MATCH-INGRESULT. Each statement is individually checked if it passes the *privacy filter*, the *confidence filter* and the *temporal filter*. The *privacy filter* checks if the statement.access is either set to public, or if it is set to friends, that the *friends relation* holds between the query.requestor and the statement.owner, or if it is set to private that the query.requestor is the same as statement.owner. Further filters can be added to this part of the algorithm.

> **procedure** *filtering* (query, MATCHINGRESULT)
>   **forall** statement ∈ MATCHINGRESULT
>   **begin**
>   **if** (statement.access = public
>   **or** (statement.access = friends
>   ∧ query.requestor ∈ *friends* (statement.owner))
>   **or** (statement.access = private
>   ∧ query.requestor = statement.owner))
>   **and** statement.confidence ≥ query.minConfidence
>   ∧ statement.confidence ≤ query.maxConfidence
>   **and** statement.start ≥ query.fromTime
>   ∧ statement.start ≤ query.untilTime
>   **then**   **add** statement **to** FILTERINGRESULT
>   **end**
>   **return** FILTERINGRESULT

As every user and every system is allowed to enter statements into repositories, some of this information might be contradictious. Conflicts among SITUATIONAL-STATEMENTS like for example a contradiction caused by different opinions of different creators or changed values over time are loosely categorized in the following listing.

1. ON THE SYNTACTICAL LEVEL: statements can for instance differ in the use of the statement attributes like subject, predicate, object, range, start etc., clear *modeling guidelines* are necessary.
2. ON THE SEMANTICAL LEVEL: the systems are not forced to use the same vocabulary, to say the same ontology, to represent the meaning of the concepts, which leads to the user model integration problem number one: *ontology merging* and *semantic web integration*.
3. ON THE OBSERVATION AND INFERENCE LEVEL: several sensors can see same things differently and claim to be right, measurement errors can occur, systems may have preferred information sources
4. ON THE TEMPORAL AND SPATIAL LEVEL: information can be out of date or out of spatial range, a degree of expiry can hold. Reasoning on temporal and spatial meta data is necessary
5. ON THE PRIVACY AND TRUST LEVEL: information can be hidden, incomplete, secret or falsified on purpose, a system of trustworthiness could be applied

Conflict Resolvers are a special kind of filters that control the *conflict resolution process*. An ordered list of these resolvers define the *conflict resolution strategy*. They are modeled in the `query.strategy` attribute. These resolvers are needed if the *match process* and *filter process* leave several conflicting statements as possible answers. Three kinds of conflict resolvers can be identified: the *most(n)*-resolvers that use meta data for their decision, the *add*-resolvers that add expired or replaced statements to the conflict sets, and the *return*-resolvers that don't use any data for their selection.

mostRecent(n). Especially where sensors send new statements on a frequent basis, values tend to change quicker as they expire. This leads to conflicting non-expired statements. The *mostRecent(n)* resolver returns the *n* newest non-expired statements, where *n* is a natural number between 1 and the number of remaining statements.

mostNamed(n). If there are many statements that claim A and only a few claim B or something else, than *n* of the "most named" statements are returned. Of course it is not sure that the majority necessarily tells the truth but it could be a reasonable rule of thumb for some cases.

mostConfident(n). If the confidence values of several conflicting statements can be compared with each other, it seems to be an obvious decision to return the *n* statements with the highest confidence value.

mostSpecific(n). If the `range` or the `object` of a statement is more specific than in others, the *n* "most specific" statements are returned by this resolver. For example if: `auxiliary=`*hasKnowledge*, `predicate=`*chess* and first `range=`*yesNo* while the second `range=`*Novice-Occasional-Professional-Expert-Grandmaster*, the statement with the second range contains a more specific information. Another specificity range ordering is for example: *yesNo* < *lowMediumHigh* < *0%-100%*

mostPersonal(n). If the `creator` of the statement is the same as the statement's `subject` (a self-reflecting statement), this statement is preferred by the *mostPersonal(n)* resolver. Furthermore, if an *is-friend-of relation* is defined, statements by friends could be preferred to statements by others. However, this resolver bears the problem, that users might not be their best judge. However due to privacy arguments, the user's own statements that are given (on purpose) should be preferred. (An alternative approach with the `creator` information could have been to define a *trusted-creator relation*.)

addExpired. Per default the already expired statements are filtered out. However, if one wants to take them into consideration, the *addExpired*-resolver adds these statements to the conflict sets.

addReplaced. Statements that are marked with the replaced-flag by other statements, are also per default filtered out and not considered in the situation retrieval process. The *addReplaced*-resolver brings these statements back into the process.

addPrivate. Statements that do not pass the privacy settings are always filtered out. However, for development, testing and administrational reasons experimental private statements may also be recognized with the *addPrivate*-resolver.

returnAll. If the remaining conflict set should not be resolved any further by the integrated mechanism, the resolver *returnAll* returns all remaining statements that can then be resolved by an external conflict resolution method, resolved by introspection or left unresolved since our approach also allows conflicting extensions in parallel.

returnNone. If there still occurs a conflict that could not be resolved until the *return-None* resolver is applied, no statement is returned. This is a very safe way not to say something wrong. This rule could be compared with *sceptical inheritance* in non-monotonic reasoning: *I don't know!*

returnRandom(n). If after applying several filters still no unique value is found but a unique answer is expected, a random pick will be offered by this resolver. This credulous behavior is selected by the requestor and therefore acceptable.

returnDialog. If no unique value is found, an alternative conflict resolution strategy could be *clarification by dialog*[4]. In some cases an appropriate human-computer dialog will be initiated in this case.

These conflict resolver rules are based on common sense heuristics. An important issue to keep in mind is the problem that resolvers and strategies imply uncertainty. To contribute to this fact, the `confidence` value of the resulting statement is appropriately changed, furthermore the conflict situation is added to the `evidence` attribute.

An example is given by the ALARMMANAGER, see [15]. It is a notification service for instrumented environments that adapts the presentation of announcements to the user's state of arousal and the user's location. Both are retrieved from the GUMO enabled u2m.org user model service. The location is derived from a positioning service application. This service runs on the user's PDA and uses infrared beacons and RFID tags that are installed in the environment to estimate the location of the user which is then sent to the user model service. Further examples and discussions can be found in [16].

6   Summary

We have introduced an integrated architecture for *Situation Modeling* and *Smart Context Retrieval*. We have clarified a model for situated interaction and context-awareness. The context exchange language `UserML` has been presented as well as the general user model & context ontology GUMO. Our approach bases on semantic web technology and a complex conflict resolution and query concept, in order to be flexible enough to support adaptation in human-computer interaction in mobile and ubiquitous computing. An important question for further research is if the analysis of the correlation between conflict resolution strategies and the user model and context ontology dimensions.

Acknowledgements

This work has been supported by the International Post-Graduate College *Language Technology and Cognitive Systems* at Saarland University and the University of Edinburgh. This research has also been supported by the German Science Foundation (DFG)

---

[4] The idea of *clarification by dialog* was recommend by Vania Dimitrova.

in its Collaborative Research Center on Resource-Adaptive Cognitive Processes (SFB 378) Project EM 4 REAL and Project EM 5 READY, as well as by the German Ministry of Education and Research (BMBF) under grant 524-40001-01 IW C03 within the project SPECTER at the German Research Center for Artificial Intelligence (DFKI).

References

1. Kray, C.: Situated Interaction on Spatial Topics. Volume 274 of DISKI. Aka Verlag, Berlin (2003) ISBN 1-58603-391-3.
2. McCarthy, J., Buvac, S.: Formalizing context (expanded notes). In Aliseda, A., van Glabbek, R.J., Westerstahl, D., eds.: Computing Natural Language. Volume 81 of CSLI Lecture Notes. Center for the Study of Language and Information, Standford University, CA (1998) 13–50
3. Dey, A.K., Abowd, G.D.: Towards a better understanding of context and context-awareness. Technical Report GIT-GVU-99-22, College of Computing, Georgia Institute of Technology, Atlanta, Georgia, U.S.A. (1999)
4. Jameson, A.: Modeling both the context and the user. Personal Technologies **5** (2001) 29–33
5. Heckmann, D.: Introducing situational statements as an integrating data structure for user modeling, context-awareness and resource-adaptive computing. In Hoto, A., Stumme, G., eds.: LLWA Lehren - Lernen - Wissen - Adaptivität (ABIS2003), Karlsruhe, Germany (2003) 283–286
6. Heckmann, D., Brandherm, B., Schmitz, M., Schwartz, T., von Wilamowitz-Moellendorf, B.M.: GUMO - the general user model ontology. In: Proceedings of the 10th International Conference on User Modeling, Edinburgh, Scotland, LNAI 3538: Springer, Berlin Heidelberg (2005) 428–432
7. Fensel, D.: Ontologies: A Silver Bullet for Knowledge Management and Electronic Commerce. Springer, Berlin Heidelberg (2001)
8. Heckmann, D., Krüger, A.: A user modeling markup language (UserML) for ubiquitous computing. In: Proceedings of the 8th International Conference on User Modeling, Johnstown, PA, USA, LNAI 2702: Springer, Berlin Heidelberg (2003) 393–397
9. Ora Lassila, R.R.S.: Resource Description Framework (RDF) Model and Syntax Specification. W3C. (1999) W3C recommendation.
10. McGuinness, D.L., van Harmelen, F.: OWL web ontology language overview. http://www.w3.org/TR/owl-features/ W3C Recommendation (2003)
11. Pease, A., Niles, I., Li, J.: The suggested upper merged ontology: A large ontology for the semanticweb and its applications. In: AAAI-2002Workshop on Ontologies and the Semantic Web. Working Notes (2002) http://projects.teknowledge.com/AAAI-2002/Pease.ps.
12. Stahl, C., Heckmann, D.: Using semantic web technology for ubiquitous location and situation modeling. The Journal of Geographic Information Sciences CPGIS: Berkeley **10** (2004) 157–165
13. Jameson, A.: Systems That Adapt to Their Users: An Integrative Perspective. Department of Computer Science, Saarland University, Saarbrücken, Germany (2001)
14. Kobsa, A.: Generic user modeling systems. User Modelling and User-Adapted Interaction Journal **11** (2001) 49–63
15. Brandherm, B., Schmitz, M.: Presentation of a modular framework for interpretation of sensor data with dynamic Bayesian networks on mobile devices. In: LWA 2004, Lernen Wissensentdeckung Adaptivität, Humboldt-Universität zu Berlin, Germany (2004) 9–10
16. Heckmann, D.: Ubiquitous User Modeling. PhD thesis, Computer Science Department, Saarland University, Germany (2005)

An Architecture for Developing Context-Aware Systems

Kaiyu Wan, Vasu Alagar, and Joey Paquet

Concordia University, Montreal, Canada
{ky_wan, alagar, paquet}@cse.concordia.ca

Abstract. This paper proposes a component-based architecture and development methodology for context-aware systems. A context is formally defined from relational point of view. The architecture of a context-aware system is conceived as a composition of the two components *context constructor* and *context adapter*. To process dynamically changing contextual information, we introduce context calculus as the formal basis of context manipulation. The information and its sources are abstracted within this formal definition. As an illustration of the principles involved in developing a context-aware system, we discuss the *Anti-lock Braking System* problem.

Keywords: context, context-awareness, co-design, components.

1 Introduction

In this paper we discuss a three-tiered formalism, a component-based architecture derived from it and a development method for constructing *context-aware* systems. As an illustration of the principles involved in developing a context-aware system, we discuss the architecture of the *Anti-lock Braking System* (ABS) problem.

Context is a rich concept and is hard to define. The meaning of "context" is tacitly understood and used by researchers in diverse disciplines, such as linguistics, AI, and HCI. Informally, many types of information, such as information on *location*, *time*, *identity of nearby objects*, and *emotional state*, are bundled together in modeling contexts. An informal definition of context, according to Dey [3] is *any information that can be used to characterize an entity*, where an entity is *a person, place, or object that is considered relevant to the interaction between user and application*. Determining the entities, and more importantly the properties to characterize them are clearly problem-dependent. In database studies, entity-relationship models are used for such a characterization. However, in such models interaction between entities cannot be shown. In object-oriented modeling of software systems, there are notations to characterize object behavior and show object iterations. It is important to adopt such notations, besides taking intuitive definitions, for developing context-aware applications. In this paper we use component notation to describe an architecture for context-aware systems.

1.1 Examples of Contexts

In natural language processing [3, 4], contexts arise as *situations* for interpreting natural language constructs. In AI, the notion of context was first introduced by McCarthy [8]

T.R. Roth-Berghofer, S. Schulz, and D.B. Leake (Eds.): MRC 2005, LNAI 3946, pp. 48–61, 2006.

and later used by Guha [6] as a means of expressing assumptions made by natural language expressions. Intensional logic [5] is a branch of mathematical logic which is used to describe precisely context-dependent entities. In *intensional programming* paradigm, which has its foundations in Intensional Logic, the real meaning of an expression, called *intension*, is a function from contexts to values, and the value of the intension at any particular context, called the *extension*, is obtained by applying context operators to the intension. In programming languages, *static* context introduces constants, definitions, and constraints, and *dynamic* context processes the executable information for evaluating expressions. In modeling human-computer interaction, the context includes the *physical place* of the user, the *time constraints and services*, and the system's knowledge of the user profile. In Ubiquitous computing [7], context is understood as both *situated* and *environmental*. The environment of an entity must be understood as a finite and relevant, but dynamically changing world. In spite of the finiteness assumption, for many applications the environment can neither be fully predetermined nor be fully characterized. Sensory units should be well calibrated to measure the environmental contextual information with minimum error.

1.2   Features of Context-Awareness

A system that interacts with the environment may not have sufficient knowledge of the environment, and it is the responsibility of the environment observation units to communicate the environment context parameters to the system. Often the systems have only limited resources, yet they are expected to function appropriately so that its services are timely, and remain useful for the user in that environment. The term "context-awareness" was first introduced by Schilit [12] to refer to such systems. Pascoe [10] gave a characterization of context-awareness as the ability of the system to *sense, interpret, and adapt* to different contexts. Hence, a context-aware system will have to uses contexts that have all the interpretations given in Section 1.1.

The distinguishing features of a context-aware system are *perception* and *adaptation*. Both the computing system and the entities interacting with and controlling the system share some knowledge of the context where the interactions and perceptions will take place. This motivates us to model contexts as *relations*. The perception feature makes the system aware of the entities in the region of its governance, and triggers context-driven interaction among the entities. The nature of interaction is in general heterogeneous, with varying levels of synchrony, concurrency, and delay. However, the system is to be fully controlled and guided by the time-varying contextual conditions and system's progress should remain both predictable and deterministic. In order to achieve determinism and predictability, the system adapts to the stimuli from its environment by learning the changing relationship among the entities and acting to fulfill the intentions expressed by the entities. That is, it reconstructs contexts based on the information it gathers from five distinguished *dimensions*, which we call W5:

- [*perception*]- *who* requires the service?
- [*interaction*]- *what* service is required?
- [*locality*]- *where* to provide the service?
- [*timeliness*]- *when* to provide the service?
- [*reasoning*]- *why* an action is required?

1.3 Contributions

The paper is organized as follows: In Section 2 we discuss the rationale for an architecture based on a three-tiered formalism and provide an overview of each tier. In Section 3 we introduce the formal description notation for the architecture. Perception and construction of primitive contexts, context management, and adaptation are done at the three tiers. A formal treatment of context management done in Tier 2 is given in Section 4. This discussion includes context definition, and context calculus. The adaptive behavior of the system (Tier 3) is discussed in Section 5. Section 5.1 gives a co-design for model perception and primitive context construction in Tier 1. The example of a context-aware system taken up for illustration is the ABS available in most of the modern day cars. We discuss ABS architecture in Section 6. We conclude the paper in Section 7 with a review of our ongoing work.

2 Three-Tiered Modeling

Context-aware systems are notoriously heterogeneous in terms of device types, context interpretations, and adaptation requirements. We introduce a three-tiered model as a solution to the complexity created by this heterogeneity problem. Perception involves the objects perceived, the devices used for observing the objects, and the observational measurements. The objects and the devices are low level abstractions belonging to one tier. The observations of the device signals are converted to symbolic forms, and composed into a mathematical representation. This description belongs to one tier, which links the low level abstraction to the high-level adaptation abstraction. Thus we end up with three tiers. The architecture that we propose encapsulates the three tiers into two components, one of which deals with context construction and the other deals with context adaptation.

Tier 1 is a description of "*see, gather, control*, and *modify*" features of perception abstractions. In describing Tier 1, we consider the environment, sensing mechanism, and functional transformation of the observed raw data. The entities in the environment seen by Tier 1 include users, programmable parts, sensors, and actuators. We emphasize that user related information is either conveyed directly by the user to one of the devices or is perceived automatically by some devices. The context surrounding the user may be users and devices. We filter away the heterogeneity in the set of entities by encapsulating the behavior of each entity separately as a *Co-design Finite State Machine* (CFSM) [2]. It describes the assembled *symbolic* representations of the observations and *notifies* Tier 2. Tier 2 *receives* the notification from Tier 1, and constructs contexts that reflect the current awareness. The formal basis of Tier 2 functionality is the context theory explained in Section 4. Tier 2 uses the *context calculus toolkit* provided by the theory to construct general contexts, de-construct and modify them. Tier 2 *notifies* current context information to Tier 3. Tier 3 *receives* current context from Tier 2 and determines how the system has to adapt itself. The modified context information is given to Tier 1 through Tier 2. The adaptive control mechanism in Tier 3 is modeled by an *Extended State Machine* (ESM) model, which has been used to model real-time reactive systems [9, 13]. Figure 1 shows the three tiers and the representations for each tier.

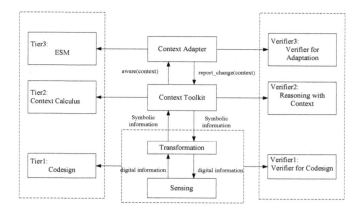

Fig. 1. The Three-Tiered Formalism of Context-Aware System

3 Architecture of the System

In this section we give a component model of the three tiers, following the terminology and notation in [11]. A component type $\mathbf{T} = \langle \mathbf{F}, \mathbf{A} \rangle$ defines a black-box view, called frame $\mathbf{F}$, and a grey-box view, called architecture $\mathbf{A}$. The type $\mathbf{T}$ may include more than one grey-box view. The black-box defines the interface types, and the particular grey-box view $\mathbf{A}$ as a structured implemented version of $\mathbf{F}$. An interface of a component is an instance of either *notifies-interface* type or *receives-interface* type. The architecture is primitive if its structuring is to be provided in an underlying implementation (outside the scope of component specification language). A non-primitive architecture includes several subcomponents *nested* to several levels. A specific implementation of a non-primitive structure is obtained by (i) instantiating adjacent level subcomponents, and (ii) specifying the interconnection between subcomponents by means of their interface ties. The four kinds of ties between the interfaces of two component instances A and B are as follows:

- *binding*($\xrightarrow{bind}$) of a receives-interface to a notifies-interface between two components at the same level of nesting (assume that all components are subcomponents of the system),
- *delegating* ($\xrightarrow{delg}$) from a receives-interface of a component A to a receives-interface of a subcomponent B of A on the adjacent level,
- *subsuming* ($\xrightarrow{subs}$) from a notifies-interface of a component B to a notifies-interface of a component A, where B is a subcomponent of A on its adjacent level, and
- *exempting* ($\xrightarrow{exem}$) an interface of a component from participating in the architectural connection.

Following the above conventions we describe the architecture construction of a context-aware system. We encapsulate Tier 1 as a component CC, which constructs contexts out of the observations. Component CC is a composition of two components: component E modeling the environment and the primitive component T modeling the transformation unit that constructs elementary contexts from the observations in the environment.

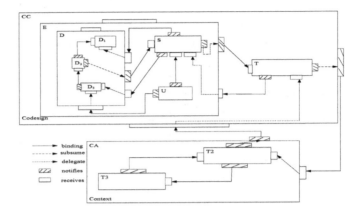

Fig. 2. The Architecture of Context-Aware System

Component E is composed from component D modeling the collection of devices, the primitive component U modeling the user, and the primitive component S modeling the sensing mechanism, which is a central unit sensing all the information the system needs. The component D is composed from subcomponents D_i, $i = 1, \ldots, k$, where each component modeling one device is primitive. We compose the primitive component T_2 corresponding to Tier 2 and the primitive component T_3 corresponding to Tier 3. The resulting component CA is the context adapter. A context-aware system is a composition of CC and CA, as shown in Figure 2.

3.1   Interface Specification

In specifying the component interfaces we use lowercase letters to name the interfaces of a primitive structure and uppercase letters to name the interfaces of a non-primitive structure. Names in bold font denote types. The instances of an interface type a are denoted as a_i, $i = 1, \ldots,$. The naming convention is as follows: (1) *A is a primitive structure*: the notifies and receives interface types for A are $a \star n$ and $a \star r$, where the place holder $\star$ is the structure name with which A might communicate; and (2) *A is a non-primitive structure*: the notifies and receives interface types for A are $A\star n$ and $A\star r$, where the place holder $\star$ is the structure name with which A might communicate. As an example of the naming convention, the name Abr is assigned to the receives-interface type of the non-primitive structure A in a composition with the notifies-interface type bAn of the primitive structure B.

In the architecture shown in Figure 2, the components for devices, user, sensing, transformation, contextor T_2 and adapter T_3 are primitive components. A device may notify another device or may notify the sensing mechanism. Likewise, a device may receive a notification from another device or from sensing mechanism S. Detecting the location, orientation, and intentions of a user are the functionalities of the devices. A user may also provide the sensing unit certain QoS parameters. As an example, the user may set the resolution factors for images delivered by a video camera device. In the model, these are acts of notification from the user to the sensing unit, which in turn iteratively interacts with the sensory devices until the received data satisfies the QoS parameters. As an example, of primitive component specification is given below: we show below the interface specification for a device:

Interface Specification	Description
interface **ddn** { *void PutEvent(out string did,* *out string event);* };	The notification from a device to another device includes the identity of the notifying device, and the content of notification.

interface **dsn** {
void PutObservation(out string did,
    out string data, out string metadata);
};
	The notification from a device to the sensing device includes the identity of the notifying device, the observation and the metadata on the observation. The metadata may include precision and metrics factors.

interface **ddr** { *void GetEvent(in string did,* *in string event);* };	A device receives the identity and signal/event notified by another device.
interface **dsr** { *void GetObservation(in string did,* *in string event, in string metadata);* };	The device identified by did receives QoS information or the correction to be applied on the observation referred by $event$.
interface **dur** { *void GetParameters(in string did,* *in string data);* };	Assuming that there is only one user, the device identified by did receives the user-centric data.

After specifying all primitive components, non-primitive components are specified. As an illustration we show below the specifications for CC. The frame specification is a black-box view of CC, and the architecture specification is a grey-box view of CC. In the latter, an instance of E and an instance of T are composed, according to the interface ties shown in **bind**, **subsume**, and **delegate** sections.

| **Frame** *ContextConstructor CC* {
    notifies: **tCCn** *CCCAn*;
    receives: **tCCr** *CCCAr*;
}; | **architecture** *ContextConstructor CC* {
    inst *Environment E, TransformingUnit T*;
    bind $T : tEr$ **to** $E : Etn$, $E : Etr$ **to** $T : tEn$;
    subsume $T : tCCn$ **to** $CCCAn$;
    delegate $CCCAr$ **to** $T : tCCr$;
}; |

4   Context Theory

In this section we give a theoretical basis for assembling and manipulating the symbolic information from Tier 1 co-design into contexts. A contract between CC and CA is that an elementary context constructed by T be in the form of a *micro-context* as explained later in the section. The component T_2 is equipped with the context toolkit, as developed below. The component T_2 can assemble or disassemble a context into micro-contexts.

4.1   Context Definitions

The context information is in general *multidimensional*, where in each dimension there exists several choices to be considered. As an example, a physical context for a braking device includes certain values (rotation per minute for wheels), situation (wheels locked or not), environmental conditions (friction on the road, gradient of road surface), and user characteristics (applying brakes or not). Here we have enumerated four dimensions. In each dimension, there are several possible ways to represent information. We say information in each dimension is *tagged*, and given the set of dimensions (dimension names), and a tag set for each dimension, then it is easy to see why the context should be defined as a of a finite union of relations. Hence, we formally introduce $DIM = \{d_1, d_2, \ldots, d_n\}$, a finite set of dimension names, and associate with each $d_i \in DIM$ a unique enumerable tag set X_i. Let $TAG = \{X_1, \ldots, X_r\}$ denote the set of tag sets. There exists functions $f_{dimtotag} : DIM \rightarrow TAG$, such that the function $f_{dimtotag}$ associates with every $d_i \in DIM$ exactly one tag X_j in TAG.

Definition 1. *Consider the relations*

$$P_i = \{d_i\} \times f_{dimtotag}(d_i) \quad 1 \le i \le n$$

A context C, given $(DIM, f_{dimtotag})$, is a finite subset of $\bigcup_{i=1}^{n} P_i$. The degree of the context C is $|\Delta|$, where $\Delta \subset DIM$ includes the dimensions that appear in C.

A finite context is written using *enumeration* syntax. The set enumeration syntax of a context C is $C = \{(d, x) \mid d \in \Delta, x \in f_{dimtotag}(d)\}$. The concrete syntax for a finite context is $[d_1 : x_1, \ldots, d_n : x_n]$, where $d_1, \ldots, d_n$ are dimension names, and x_i is the tag for dimension d_i.

We say a context C is *simple* (s_context), if $(d_i, x_i), (d_j, x_j) \in C \Rightarrow d_i \ne d_j$. A simple context C of degree 1 is called a *micro* (m_context) context. Several functions on contexts are predefined. Two of the basic functions are dim and tag. They respectively extract the set of dimensions and the associated tag values from a set of contexts.

In general, a set of contexts may include contexts of different degrees. We use the syntax $Box[\Delta \mid p]$ to introduce a finite set of contexts in which all contexts are defined over $\Delta \subseteq DIM$, have the same degree $|\Delta|$, and the tags in every context satisfy the predicate p.

Definition 2. *Let $\Delta = \{d_1, \ldots, d_k\}$, where $d_i \in DIM$, $i = 1, \ldots, k$, and p is a k-ary predicate defined on the tuples of the relation $\Pi_{d \in \Delta} f_{dimtotag}(d)$. The syntax*

$$Box[\Delta \mid p] = \{s \mid s = [d_{i_1} : x_{i_1}, \ldots, d_{i_k} : x_{i_k}]\},$$

where the tuple $(x_1, \ldots, x_k)$, $x_i \in f_{dimtotag}(d_i)$, $i = 1, \ldots k$ satisfy the predicate p, introduces a set S of contexts of degree k. For each context $s \in S$ the values in $tag(s)$ satisfy the predicate p.

For example, the set of contexts defined by $Box[X, U \mid \frac{x}{4} + \frac{u}{5} \le 1 \wedge (2 \le u \le 3)]$, where $f_{dimtotag}(X) = f_{dimtotag}(U) = \mathbb{N}$ is the set of contexts $\{[X : 0, U : 2], [X : 0, U : 3], [X : 1, U : 2], [X : 1, U : 3], [X : 2, U : 2]\}$. A non-simple context is a short-hand notation for a set of simple contexts, that may not be expressible in Box notation. The context $C_4 = [X : 3, X : 4, Y : 3, Y : 2, U : blue]$ can not be expressed in Box notation. However, C_4 should be understood as the set of simple contexts $\{[X : 3, Y : 3, U : blue], [X : 3, Y : 2, U : blue], [X : 4, Y : 3, U : blue], [X : 4, Y : 2, U : blue]\}$.

4.2 Context Calculus

In our previous papers [1, 13], we have formally defined the following context operators: the *override* $\oplus$ is similar to function override; *difference* $\ominus$, *comparison* $=$, *conjunction* $\sqcap$, and *disjunction* $\sqcup$ are similar to set operators; *projection* $\downarrow$ and *hiding* $\uparrow$ are selection operators; *constructor* $[\_ : \_]$ is used to construct an atomic context; *substitution* $/$ is used to substitute values for selected tags in a context. They have formal definitions [14]. The following table shows the formal syntax for well-formed context expression C, and precedence rules for context operators [14].

syntax	precedence
$C ::= c \qquad\mid C = C$ $\mid\ C \supseteq C \mid C \subseteq C$ $\mid\ C \mid C \quad\mid C/C$ $\mid\ C \oplus C \mid C \ominus C$ $\mid\ C \sqcap C \mid C \sqcup C$ $\mid\ C \rightleftharpoons C \mid C \rightharpoonup C$ $\mid\ C \downarrow D \mid C \uparrow D$	1. $\downarrow, \uparrow, /$ 2. $\mid$ 3. $\sqcap, \sqcup$ 4. $\oplus, \ominus$ 5. $\rightleftharpoons, \rightharpoonup$ 6. $=, \subseteq, \supseteq$

Many of the context operators can be lifted in a natural manner to *Boxes*. As an example, for a box $B = Box[\Delta \mid p]$ and context c, the expression $B \oplus c$ is the set of contexts $\{c' = \{c_i \oplus c \mid c_i \in B\}\}$. In addition to lifted operators, we have also defined [13] the operators *join* $\boxtimes$, *union* $\boxplus$, and *intersection* $\boxdot$. These three *Box* operators have equal precedence and have semantics analogous to relational algebra operators. A *Box* expression is defined as: $B ::= b \mid B \boxtimes B \mid B \boxplus B \mid B \boxdot B$.

Example 1. *The evaluation steps of the well-formed context expression $c_3 \uparrow D \oplus c_1 \mid c_2$, where $c_1 = [x : 3, y : 4, z : 5]$, $c_2 = [y : 5]$, and $c_3 = [x : 5, y : 6, w : 5]$, $D = \{w\}$, are as follows:*

1. $c_3 \uparrow D = [x : 5, y : 6]$ [Step1]
2. $c_1 \mid c_2 = c_1$ *or* c_2 *chosen nondeterministically*   [Step2]
3. Suppose in Step2, c_1 is chosen,
    $c_3 \uparrow D \oplus c_1 = [x : 3, y : 4, z : 5]$ [Step3a]
    else if c_2 is chosen,
    $c_3 \uparrow D \oplus c_2 = [x : 5, y : 5]$ [Step3b]

5 Context Construction and Adaptation

The context construction is done by component CC, which is made up of device components and user models. The context adaptation component CA is made up of two subcomponents T_2 and T_3. In this section we explain the behavior of CC and CA.

5.1 Behavior of CC

The behavior of each primitive component in CC is modeled by a CFSM. The semantics of a transition in a CFSM is that there exists an event occurrence that satisfies the transition specification: (For each state s_i, let $\mathcal{E}(s_i)$ denote the set of events that are possible in s_i.)

$$\frac{(s_i, v_i) \; \wedge \; e \; \in \mathcal{E}(s_i) \; \wedge \; p[(v_i)] \; \wedge \; tc(t)}{(s_i, v_i) \; \xrightarrow{e} \; (s_j, v_j) \; \wedge \; q[(v_j)] \; \wedge \; Init(t)}$$

where, (1) e is the event triggering the transition, (2) $g = p \wedge tc$, where p, a predicate on the variables in pre-state and tc is the time constraint predicate, and (3) $a = q \wedge Init$, where q is a predicate on the variables in the post-state, and $Init$ is timer initialization predicate. Elementary contexts are constructed by the CFSM of the component T, after the CFSMs corresponding to the components in E collaborate among themselves. Due to the reactive property of the CFSMs, contexts are constructed during every cycle of getting new information from the devices.

5.2 Behavior of CA

The component T_2 is fully described by the context theory. The behavior model of T_3 is an ESM, which has a finite number of states with transitions between states labeled by events triggering the transition. At each state, the ESM may receive context information composed by T_2, query the context database and/or compute a mathematical function, and output the results to CC. In systems involving continuous dynamics, adaptation is based on the result of evaluation of a mathematical function available in the application domain; in discrete systems involving a multitude of dimensions, a search engine operates on a context database to generate context-dependent actions, which implicitly change the context. The states may include auxiliary data variables for local computations. The state transition semantics is

$$\frac{s_i \; \wedge \; e \; \in \mathcal{E}(s_i) \; \wedge \; var_g(s_i) \; \wedge \; con_g(s_i) \; \wedge \; tc(t)}{s_i \; \xrightarrow{e} \; s_j \; \wedge \; var_a(s_j) \; \wedge \; rtc(t)}$$

where (1) var_g is a predicate in Conjunctive Normal Form, in which each atomic predicate expresses a constraint on one variable in state s_i; (2) con_g is also a predicate in Conjunctive Normal Form, in which each atomic predicate expresses a constraint on context information in state s_i; and (3) tc is a conjunction of linear time predicates of the form $Lower \leq t \leq Upper$, where t is the valuation of a local clock. An execution in the ESM is a sequence of transitions starting from an initial state. The behavior of the ESM is the set of executions.

Context-Dependent Adaptation. Adaptation mechanism uses context-dependent function definitions as well as a database of contexts constructed on the five dimensions W5. A context-dependent expression E is a pair $(\lambda \cdot E, \mu \cdot E)$, where $\lambda \cdot E = \{E_1, \ldots, E_k\}$, is the set of definitions for E and $\mu \cdot E = \{\omega_1, \ldots, \omega_k\}$ is the set of regions corresponding to the definitions. In our context theory a region has a Box representation. For any arbitrary context c, $E@c = \{E_1@c \sqcap \omega_1, \ldots, E_k@c \sqcap \omega_k\}$. An immediate consequence is that contexts can be used as parameters in a function definition. Let $f : X \times Y \times Z \times C \to W$, where C is a set of contexts, and $f(x, y, z, c), x \in X, y \in Y, z \in Z, c \in C$, be defined such that for different context values, the function's definitions are different. In Example 2, we define a context-dependent function

$f(x, y, z, c)$ corresponding to different context regions shared by the upper halves of a sphere and a cone.

Example 2.

> $f(x,y,z,c)= (\lambda \cdot f, \mu \cdot f)$, where
> $\lambda \cdot f = \{2x^3 + y - 6, x + y^2, z^3 + y\}$,
> $\mu \cdot f = \{B_1 \ominus B_2, B_2 \ominus B_1, B_1 \sqcap B_2\}$,
> $B_1 = Box[X, Y, Z \mid x^2 + z^2 \leq 16 \wedge x = \frac{1}{2}z \wedge z \geq 0]$,
> $B_2 = Box[X, Y, Z \mid x^2 + y^2 + z^2 \leq 9 \wedge z \geq 0]$,
> *end*

In addition, one or more *Box*es having the five dimensions of W5, mentioned in Section 1, with a set of constraints on their tag values can be interpreted as a region of lattice points or a multi-dimensional relation. These relations, stored in a database, may also be used for context adaptation. With each relation in the database an adaptation information is associated. Whenever a context in a relation is satisfied, the adaptation rule associated with it is fired. Search queries can use context expressions to search the database and determine the adaptive action.

6   Case Study - Anti-lock Braking System (ABS)

ABS is a typical example of a context-aware system [7], in which both situated and environmental information should become part of a context. In addition, time constraints may also be included, either as efficiency requirements or time-dependent functionality requirements. It *senses* the driver's situation (braking or not), the status of the car's wheel (locked or rotating), the road conditions and other environmental contexts. Once it detects the locking of a wheel it reduces the braking force repeatedly until the wheel starts rotating again. Our design is more general and formal than the one suggested in [7].

The functionalities of the ABS components and assumptions on them are as follows:

1. *Functionalities*
   (a) Wheel sensors sense the speed of the wheels, and inform the controller when the wheel is locked due to too much pressure.
   (b) Braking units sense the amount of pressure exerted by the driver and inform the pressure intensity to the controller. They also control the pressure in the brake lines of the vehicle.
   (c) The controller receives the environmental contexts, such as road friction, brake torque and external disturbances, and calculates the amount of pressure that the brake should actually release.
2. *Assumptions*
   (a) The *pumping* of the brake pressure, the cycle of receiving current context and calculating a new context for adaptation, happens 20 or more times per second.
   (b) Every 20 time units, the sensors and brakes report their information when wheels work properly. However, once wheels are lock up, the sensors and brakes report their information every 1 time unit.
   (c) Only when all the wheels resume rotating, the controller stops the pumping, and all the components transfer to *normal* states.

6.1 The ABS Model

ABS has two types of devices, wheels and braking units. There are four wheels and braking units in a car, but they share the same functionalities. Hence, for simplicity, we only draw one wheel (D_{11}) and one brake unit (D_{21}) in the architecture. Environmental contexts such as the road friction, the brake torque, the tire pressure, and the current wheel speed are sensed by the Sensing mechanism S. They are converted into symbolic form by the transformer and passed as micro_contexts to T_2. Applying the $\oplus$ operation to micro_contexts, a simple context c is constructed and passed as a parameter to T_3. The architecture of ABS is shown in Figure 3. The frame specification, architecture specification and the architecture protocol of ABS can be defined following the specification shown in Section 3. We skip the details.

Device wheel and braking units, which communicate with and controlled by the adaptation unit, are modeled as CFSMs, as shown in Figure 4, with time constraints. The formal behavior model of context adaptation is shown in Figure 5. In the state *normalWork* it receives context information every 20 times units and puts it into its context queue. If the message *Locked(id)* is received, it enters into *urgency* state. In

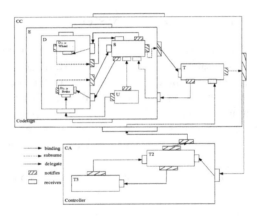

Fig. 3. The Architecture of ABS

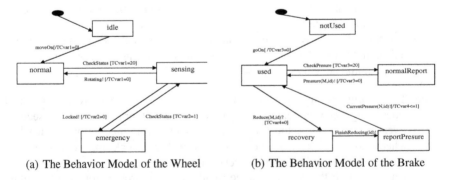

(a) The Behavior Model of the Wheel   (b) The Behavior Model of the Brake

Fig. 4. Formal Behavior Model of Context Constructor of ABS

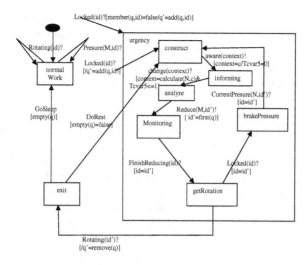

Fig. 5. Formal Behavior Model of Context Adapter of ABS

that state, the adaptation mechanism computes a new context within 1 time unit, and analyzes the new context using the provided context operator: projection($\downarrow$). This calculation determines the amount of pressure to be reduced in the new context. The reduced pressure is notified to the brake unit. After the brake notifies the adapter that the reduced pressure has been applied, the adapter notifies the wheel to test whether or not the wheel is locked. If the wheel notifies the adapter that the wheel is rotating, from *urgency* state the adapter changes to *exit* state. However, if the wheel is still locked, the adapter waits for the brake to report the current pressure within 1 time unit, enters into *construct* state, and the whole construct-reduce-apply-test procedure is repeated until the wheel starts rotating. At the *exit* state, if no other wheel is locked, the adapter changes to *normalWork* state; otherwise, it repeats the cycle of actions to deal with another locked wheel. The function $calculate(p, c)$ is context dependent and its evaluation results in a new context c' is notified to the devices.

7 Conclusion

The main contribution of this paper is a component-based architecture for developing context-aware systems. The architecture is based on a three-tiered model, where the first tier deals with perception, the second tier deals with context management, and the third tier deals with context adaptation. These tiers are wrapped into two interacting components *CC*(context constructor) and *CA* (context adapter). A context-aware system is realized as a composition of *CC* and *CA*.

The significant aspects of the architecture are: (1) *modularity:* each component is fully encapsulated, and their relative independence promotes easy refinements and maximal reuse; (2) *co-design:* the sensing components in CC require hardware implementation, and the transformer T requires a software implementation, and the tight coupling between them can be optimized through appropriate hardware/software partitioning; (3)

context toolkit: component T_2 implements context operators and provides precedence rules, and their implementation is quite independent of the nature of contexts (supplied by CC) and adaptation (at T_3); and (4) *knowledge base:* T_3 may be given the knowledge base that is specific to an application, and hence the adaptation/controller can be reused for different applications by simply rebooting a new knowledge base for every new application.

We are continuing our research in two directions. We have shown [1, 13] that the intensional programming language Lucx has the expressive power to program agent communication and real-time reactive systems. Lucx, in which contexts are first class objects, can use them to program CFSMs and ESMs. Hence it is a natural choice for programming context-aware system as described in this paper. We are currently working on the details of implementing context-aware systems in Lucx. The second aspect of our ongoing research is concerned with reasoning and verification. Since context-aware systems are reactive and adaptive to changing environmental situations, proving the current functioning of the system depends on the accuracy of sensory devices. Sensory devices produce signals that are converted to symbolic forms. Necessarily, this process is error-prone, and consequently not all context information will be accurately quantified. This raises an important distinction between correctness issues addressed for synchronous languages and the CFSMs used in context-aware applications: that some of the CFSMs are required to *auto-evaluate* the quality of service it provides. With every data generated the device exports a meta-data to express the quality of the value, including the metric used for its measurement. In order to guarantee that the design satisfies a desired property, the design methodology should be verification-driven. The verification tool uses the meta-data as an aid to refine the specification of the system. This aspect of verification-driven design and implementation of context-aware systems is an integral part of our ongoing research.

Acknowledgment. This work is supported by grants from the Natural Sciences and Engineering Research Council of Canada.

References

1. V. S. Alagar, J. Paquet, K. Wan. *Intensional Programming for Agent Communication.* Proceedings of DALT'04, New York, July 2004, Lecture Notes in Computer Science, Springer-Verlag, Vol. 3476, Page 239-255.
2. F. Balarin, E. Sentovich, M. Chiodo, P. Giusto, H. Hsieh, B. Tabbara, A. Jurecska, L. Lavagno, C. Passerone, K. Suzuki, and A. Sangiovanni-Vincentelli. *Hardware-Software Co-design of Embedded Systems – The POLIS approach.* Kluwer Academic Publishers, 1997.
3. A.K.Dey. *Understanding and Using Context.* Personal and Ubiquitous Computing Journal 5(1).pp.4-7.2001.
4. A.K. Dey, D. Salber, G.D. Abowd. *A Conceptual Framework and a Toolkit for Supporting the Rapid Prototyping of Context-aware Applications*, anchor article of a special issue on Human Computer Interaction, Vol.16, (2001).
5. D.Dowty, R.Wall, and S.Peters. *Introduction to Montague Semantics.* Reidel Publishing Company, 1981.
6. R. V. Guha. *Contexts: A Formalization and Some Applications.* Ph.d thesis, Stanford University, February 10,1995.

7. Cheverst, K., N. Davies, K. Mitchell and C. Efstratiou. *Using Context as a Crystal Ball: Rewards and Pitfalls.* Personal Technologies Journal, Vol. 3 No5, pp. 8-11, 2001.

8. J.McCarthy. *Notes on formalizing context.* In Proceedings of the Thirteenth International Joint Conference on Artificial Intelligence, 1993.

9. D. Muthiayen. *Real-Time Reactive System Development - A Formal Approach Based on UML and PVS.* Phd. Thesis, Department of Computer Science, Concordia University, Montreal, Canada, January 2000

10. J. Pascoe. *Adding Generic Contextual Capabilities to Wearable Computers.* In Proceedings of the 2nd International Symposium on Wearable Computers, pp. 92-99, 1998.

11. F. Plasil, S. Visnovsky. *Behavior Protocols for Software Components.* IEEE Transactions on Software Engineering archive, 28(11), pp.1056 - 1076,2002.

12. B. Schilit, M. Theimer. *Context-aware Computing Applications.* In Proceedings of the 1st International Workshop on Mobile Computing Systems Applications, pp. 85-90, 1994.

13. K. Wan, V.S. Alagar, J. Paquet. *Real Time Reactive Programming Enriched with Context.* ICTAC2004, September 2004, Lecture Notes in Computer Science, Springer-Verlag, Vol. 3407, Page 387-402.

14. K.Wan. *Lucx: Lucid Enriched with Context.* Ph.d thesis, Concordia University, January,2006.

Steps Towards Making Contextualized Decisions: How to Do What You Can, with What You Have, Where You Are

Oana Bucur, Philippe Beaune, and Olivier Boissier

Centre G2I/SMA, Ecole NS des Mines de Saint-Etienne,
158 Cours Fauriel, Saint-Etienne Cedex 2, F-42023, France
{bucur, beaune, boissier}@emse.fr

Abstract. Applications need facilities for recognizing and adapting to context in order to provide useful and user-centered results. There are several problems to be addressed when building context-aware applications, two of which being how to *define and manage* all available contextual information and how to *distinguish relevant* from non-relevant context for a given task. In this paper, we focus on the second problem and propose a context definition and model for a context-aware agent. We exploit this model to build agents that learn to select relevant context and to use it to make decisions.

1 Introduction

The rise of pervasive computing has stressed the importance of *"context"*. As defined in [7], this concept can be seen as "any information that can be used to characterize the situation of an entity", where an entity can be "a person, place or object that is considered relevant to the interaction between a user and an application, including the user and the application themselves". This definition underlines two of the major problems in current context-aware applications: the *amount of contextual information* that must be gathered, understood and used for a certain task (*"any* information [...] used to characterize the situation" gives an idea about how difficult is to define and manage all that information); and the necessity to measure some sort of *relevance* ("...information that can be *used to*...", "...that is considered *relevant*..."): even though the definition does not specify how to choose among all available context information the one that is relevant, it underlines the importance of taking into account only relevant issues.

Most of the existing works handle these problems at the design phase (some examples are [30], [9], [24], [2], etc.): the system's designer is the one that defines what information will be part of the application's *context* and makes the choice of what will be considered as relevant in that precise application. As the amount of available information is usually very large, the designer has to define at design time what concepts will be used in that system, and then make an a-priori choice of which ones are relevant.

This kind of choice about what is relevant cannot be used anymore, not in current applications. Classical applications had clearly defined boundaries, a specific purpose

T.R. Roth-Berghofer, S. Schulz, and D.B. Leake (Eds.): MRC 2005, LNAI 3946, pp. 62–85, 2006.

and a specific context of use that didn't change very much and was easily predictable. The new generation of pervasive and ubiquitous computing applications are now situated in a dynamic environment, where it is more difficult (if not impossible) to predict all changes and configurations that might occur. Considering this, a static approach to the environment (and to what might be seen as *relevant* information for a given application, in general) is not appropriate for 'new generation' applications. The designer cannot predict all possibilities, so the need for applications to deal with dynamic and unpredictable environments must be considered. We understand by this dynamical approach the fact that the application has no fixed a-priori about the environment in which it might be running and about the situations and tasks with which it might need to deal. Nonetheless, we expect the application to continue running with as few errors as possible and adapt at runtime to new circumstances and tasks.

We underline the fact that all this "dynamicity" could be useful only in the case of applications that need continuous adaptability and which have to deal with highly dynamical (and even unknown) environments. Applications that have limited and predictable functionalities, that do not need on-the-move adaptations to unknown situations and tasks, might not benefit from this approach.

In this paper, our goal is to draw a common base for dynamic adaptation in context-aware applications, by alleviating the system's designer's task (transferring it to the user or the application itself), especially in what concerns the choice of relevant information. We implement this in a context-aware multi-agent application. Context-aware multi-agent applications are composed of several heterogeneous and situated agents. Agents are able to share amongst themselves the way they use context knowledge in solving similar problems.

We detail the architectural construction of such a system and illustrate it with a case study of an open and interoperable context-aware agenda management using Multi-Agent technologies. The resulting MAS is made of several meeting schedulers agents called mySAM (my Smart Agenda Manager). mySAM assists its user in fixing meetings by negotiating them with other mySAMs and by using context knowledge to decide whether to accept or reject a meeting proposal made by another agent. Knowledge about how to select relevant context and how to use it to deal with a meeting proposal is acquired through individual and multi-agent learning.

In what follows, we detail our motivation and approach (section 2), give some brief definitions of what we call "context" (section 3), then present an architecture context-aware applications, with details on representing context, managing it and reasoning with context (sections 4 and 5). We describe the implementation and results obtained in section 6. We will then situate our approach in related work (section 7), make a short informal discussion on what are the benefits of this approach and what might be the problems (section 8), conclude and give some perspectives on our future work (section 9).

2 Motivation and Approach

As we mentioned before, the focus of our work is the question of *relevance*, in all its interesting points: how to define what is relevant, how to choose between relevant and non-relevant information and how to use relevant information when making decisions.

We think that there are at least two issues to be addressed: to have a context management as independent from the application as possible, and therefore reusable between applications; and to build applications that are not static, meaning that they should be able to accomplish different tasks in unpredictable situations. We consider that the first steps towards solving these issues are to make a *distinction between managing context and reasoning using* context to make decisions; and to *dynamically adapt the selection* of relevant information, according to the task at hand.

Motivations concerning these aspects were already explored in previous section and some more arguments will follow all along this paper.

Therefore, what we focus on is: an explicit representation of context, a management engine for everything that can be considered as context, and adding reasoning capabilities to agents in order to dynamically adapt the selection of relevant context and the decision making process.

Managing context vs. Context-based reasoning for making decisions
Taking into account these problems, we define a layered architecture. In this architecture (the overview of which is presented in *Figure 1*) we make a clear distinction between managing context and reasoning using context.

In our architecture, the context management task is not done by the agents themselves. Instead, they interact with a context-manager layer that is able to answer context-related queries. We thus disconnect all that is related to context acquisition and management (which will be done by the local context managers) from reasoning with context knowledge (done by the agents). The system architecture is structured along three layers: context sources (layer 0), context management (layer 1), a layer of agents that reason with context (layer 2). Over the last layer are dedicated applications

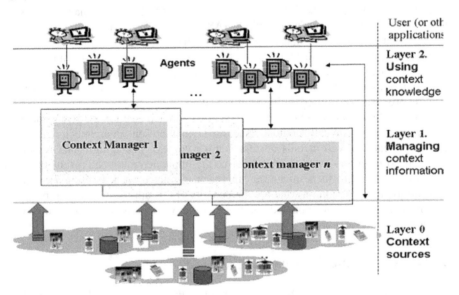

Fig. 1. Global architecture for Context-aware applications

which use agents' abilities to reason with context, or in which context can be directly used by the application itself.

Mostly, when context is used to make behaviors context-adaptable, it is used in an ad-hoc manner, without trying to propose an approach suitable for other kind of applications. However, there is some recent research in proposing a general architecture on context-aware applications, like CoBrA(Context Broker Architecture), proposed by Chen et al.[2], Socam, by Gu et al [28] or the work of Coutaz et al. [5]. The architecture we present in *Figure 1* is quite similar to all these, without focusing on how to acquire certain information from heterogeneous sources and abstract it to higher level information, but on how to represent it in a semantic and generic manner and how to reason on context knowledge based on this representation. We discuss these approaches in more details in Related Work (see Section 7).

Selecting relevant context

The main problem in context-aware applications is how to select (choose) *relevant* from non-relevant information for a specific task. The aim is to have this task (of selecting relevant information) done by the application itself, at runtime, and not in advance, by the designer. Choosing in advance what might be relevant for an application might not give expected results, as there are applications (especially highly dynamic ones) in which such a prediction of future states of the environment cannot be done. We consider that changes in environment are so unpredictable in context-aware applications, that such an a-priori choice of what is and it is not relevant is not adequate. The solution, as we see it, is to make this selection at runtime, and, if at all possible, improve the selection mechanism at each occasion.

P. Maes [16] argues that applications must deal with two main issues: *Competence* – how the application (in this case represented by a personal agent) is acquiring the knowledge needed to decide and *Trust* – what is the user's comfort in delegating tasks to those kind of agents. She describes two classical approaches in dealing with these issues: making the end-user program the interface agent, which solves the trust problem; and a "knowledge-based approach" – endowing an interface agent with extensive domain-specific background knowledge about the application and the user. This one does not solve any of the two issues completely, as the designer will have to make a significant effort to endow the agent with the extensive knowledge of the domain and the user will still not know how the agents function, so it will (probably) not trust her. The solution, as seen by Maes, is learning.

We follow this direction of using learning (therefore a dynamic adaptation to changes in environment, in tasks and user's preferences), but explicit this learning in a specific application field: the one of context-aware applications. The focus in classical applications was to have an agent that was better suited to user's needs, so the focus of learning was on agents' reasoning abilities and behavior adaptation. Our agents have a component that reason to accomplish a certain task, but we added a component that chooses relevant information to be used in this reasoning process. We argue that the step of choosing what is relevant from all available information is a very difficult and critical process for a context-aware agent, therefore our agents will use learning for two purposes. They will learn to acquire missing information about: (i) what is relevant for the task at hand and (ii) how to accomplish that task using only relevant information.

3   Basic Definitions and Our Model of "Context"

Having drawn the overall picture of our system, let us now go into detail: first by describing what we understand by "context", and then by explaining how context is represented and used to design and implement our agenda management case study. Before concluding, we will compare our approach to related work.

3.1   Basic Definitions

Agents are situated entities, meaning that they sense their environment and act accordingly. In this section we define the basic components of what we call "context": the rationale (or the finality) of its use, the elements that are part of it and the relevance of those elements given that rationale.

Finality
Going back to Dey's definition, we can define context as a cluster of information relevant for a specific purpose. Before going further, let us explain what we consider as purpose (or finality, how we will call it further on).

The *finality*, f, is the goal for which the context is used at a given moment, the focus of the activity at hand. A *finality* is the information that is the most interesting for the application at a given moment, for example: deciding what to do with an offer, explaining an action, understanding a conversation, etc. All of these are finalities that determine the way the application will consider context and act upon it, therefore compel the activity of that application.

It is the designer of the system (or the user himself) who defines the finalities that will need to be fulfilled with the use of context. For example, in an agenda-management application, finalities can be: answer to a meeting proposal, cancel a meeting, negotiate a meeting, etc. In a tourist guide application, finalities could be: propose sightseeing tour, find a museum/church/restaurant/..., etc. The set of finalities (goals to be fulfilled) is defined in association with the application at hand, as this set is the one that will guide the selection of what is relevant and what is not for that specific task.

Context attribute
As mentioned before, context is defined by a set of attributes that are relevant in a given situation. In what follows we will define the general structure of a context attribute. Usually, in literature, an element of context is either seen as a simple "label" that will have an associated value (like in [29], [1], [24], [8]), or as a more complex structure, that also includes the specification of the behavior or action to be followed when a certain value is reached (like in [9], [30]).

What we intend is to develop a specific structure within which we will be able to define a context attribute. This structure specifies the pattern of a context attribute, and does not point out which conduct should be adopted for a specific value of the context attribute. We consider a context attribute much more complex than just an attribute-value structure. We extended the simple 'attribute-value' approach to cope with the necessity of describing context information, no matter how complex it might be. What we present here is just a beginning for the complex structure that an attribute must have, because there is at least one very important aspect that we did not cope with so far, the one of possible dependencies, relationships between attributes. We plan to explore this issue in future work. In what follows, we define the elements that compose our context attribute structure for the time being.

A *context attribute (a)* designates the information defining one element of context, e.g. "ActivityLocation", "NamePerson", "ActivityDuration". Each context attribute has at least one value at a given moment, the value depending on several entities to which the attribute relates. An entity is an instance of a "person, object or place" (as mentioned in [7]), but can also be an activity, an organizational concept (role, group, another agent, etc).

For instance, the context attribute "DevicesAvailableInRoom" defines the devices that are located in a room. When trying to instantiate this attribute, the needed parameter will be the specific room that interests us. A "PersonIsMemberOf" context attribute will take a person as input, and will return (possibly) several groups in which that person takes part.

We can therefore associate to each context attribute an instantiation function called ***valueOf***. We define *valueOf* for a context attribute *a* as being the function that for the set of parameters (p_a), it calculates the value of *a* at time *t*, taken that set of parameters. There exist some context attributes (as "PersonIsMemberOf", "Supervises", "Devices-AvailableInRoom", etc.) that can take not just one, but a set of values.

Relevance

Not all attributes are relevant for a finality. We define ***isRelevant(a,f)***, a predicate stating that the attribute *a* is relevant for the finality *f*. Let us call ***RAS**(f)* the subset of **A** which defines the *Relevant Attribute Set* for the finality *f*:

$$RAS(f) = \{ \ a \in A \ | \ isRelevant(a,f)=\text{true} \ \}.$$

We call an *instantiation of context attribute* $a \in A$, the pair (a,v) where v is the set of values of *a* at a given moment. For instance, (DayOfMonth, {14}), (roleOfPersonInGroup, {Team Manager}), (PersonIsMemberOf, {MAS Group, Center_X, University_Y}) are all instantiations of the respective context attributes: *DayOfMonth, roleOfPersonInGroup, PersonIsMemberOf*. We call *I* the set of instantiated context attributes as

$$I = \{(a,v) \ | \ a \in A \ \wedge \ \textbf{\textit{valueOf}}(a,p_a)=v\}.$$

We call *Instantiated Relevant Attribute Set* of a finality *f*, ***IRAS(f)***, the set of instantiated context attributes relevant to finality *f*:

$$IRAS(f) = \{(a,v) \ | \ a \in \textbf{\textit{RAS}}(f) \ \wedge \ (a,v) \in \textbf{\textit{I}}\}.$$

Let us notice that in related work ([15], [25], [26]), the notion of "context" is often understood as being what we called ***IRAS***. To explain the difference between ***RAS*** and ***IRAS*** and the reason to make this distinction, ***Let us*** consider the following example. Given the finality f = "deciding whether to accept or not a meeting", ***RAS***(f)={"RoleOfPersonInGroup", "ActivityScheduledInSlot"} is considered, i.e. role played by the person with whom I al negotiating the meeting and if I have something already planned for that time slot. The resulting ***IRAS*** for a student may be ***IRAS***$_{\text{student}}$(f)={(RoleOfPersonInGroup, {teacher}), (ActivityScheduledInSlot, {Activity001})} and for a teacher ***IRAS***$_{\text{teacher}}$(f)={ (RoleOfPersonInGroup, {student}), (ActivityScheduledInSlot, {Activity255}). As we can see, the difference between ***IRAS*** of student and teacher may lead to different rational decisions. We think that, at least in some organizational settings (like universities, enterprises), **RAS** sets present some similarities for different users when needed to make decisions (for the same finality), but the decision itself is **IRAS**-dependent. This is just a hypothesis that needs to be further analyzed.

Taking into account the definitions that we proposed so far we are now able to describe the representation of a context attribute.

3.2 Representing Context

The problem with the context modeling is that, usually, the context-management task is very much connected to the reasoning-with-context one. We see this as a problem because, due to this type of management, the context needs to be re-defined and managed in a different manner for each application. There are several works that also argue the need to separate the managing of context from reasoning with context to determine the behavior to adopt in each situation ([2], [28], etc.). In this way, the management of context can be reused between applications, as there is just the reasoning engine (the utilization of this contextual knowledge to make decisions, plans, assist the user, etc.) that can vary from one application to another. Therefore, the distinction between the *context-management* task and the *reasoning-with-context* task imposes a clear separation between the *definition* of a context attribute and the *specification of the behavior* to be adopted in certain situations.

Given previous definitions, our aim is to represent context attributes in a general and suitable manner for any kind of applications that need to represent and reason about them. What we focus on is to have a complex representation for a context attribute, representation that will assure in this way the interoperability, but also the possibility to represent no matter what context attribute, not just the simple ones.

Several representations of context exist: contextual graphs ([1]), XML (used to define ConteXtML [24]), contextors ([4]) or object oriented models ([9]). All these representations have strengths and weaknesses. As stated in [10], lack of generality is the most frequent weakness: some representations are suited for a type of application and express a particular vision on context. There is also a lack of formal bases necessary to capture context in a consistent manner and to support reasoning on its different properties. A tentative answer in [10] was the entity-association-attribute model, which is an extension of the "attribute-value" representation, where contextual information are structured around an entity, every entity representing a physical or conceptual object. We based our proposal on this idea and on ontology as the explicit way to represent it. To take into account the need for generality and also considering the fact that we aim at having several MAS, each dealing with different contexts (that we will need to correlate in some way), an ontology-based representation seems reasonable. This is not a novel idea, the work done by Chen *et al.* ([2]) being just one example of defining context ontologies using OWL([19]).

Usually, when context is represented using ontologies, what is said to be "context" are the properties associated to each concept (or entity, how we will call it further on). For example, when defining a meeting ontology, we can define the status of a meeting as a property (*"statusMeeting"*) of the entity Meeting. This property can also be considered a context attribute. This representation is simple and does not allow more complex context attributes to be represented in the same manner. For example, if we want to define an attribute that will have the value "true" if two meetings are taking place in the same time, we need to define this attribute not as a property (as it will be connected to two meeting entities – in order to check when they take place), but as a class in itself. This approach will confer a rather heterogeneous representation of context attributes, as properties, classes, and instances. Our need for a generic representation of *all* context attributes made us propose a slightly different approach.

We define two ontologies: ***domain ontology*** and ***context ontology***. The domain ontology is used to define all concepts that will be used in the current system. The context ontology is the one that contains all context attributes that will be managed by a context manager. Let us detail first the domain ontology and then, how we build the context ontology using the concepts defined beforehand.

Domain ontology

In our domain ontology, we define a class "#Entity" as the super class of all concepts, e.g. in our agenda-management case study, #Person, #Group, #Room, #Activity, etc. are subclasses of #Entity.

For example, in Fig. 2, we show how concepts as: #Event, #Room, #Role, #Person, #Group relate to one another and how all of them are sub-classes of the root of the domain ontology: the class #Entity. Classes as #Role, #Person and #Group are subclasses of the concept #Social. There are several other properties for each concept than just *name* and *id*, as, for example, for a #Person, we defined properties as: *interests, currentActivity, isMemberOf*, etc.

Below you can see the OWL definition of the class #Person and #Social.

```
<owl:Class rdf:ID="#Person">
   <rdfs:subClassOf>
     <owl:Class rdf:ID="#Social"/>
   </rdfs:subClassOf>
  </owl:Class>
<owl:Class rdf:about="#Social">
   <rdfs:subClassOf rdf:resource="#Entity"/>
   </owl:Class>
```

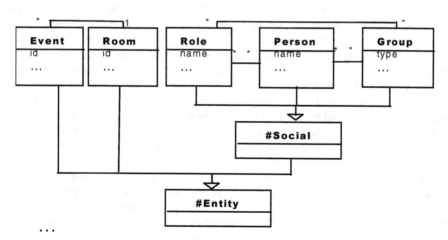

Fig. 2. A partial view on the concepts defined in a domain ontology

Using the concepts defined in this "domain ontology", we can further explain how we represented context in a different ontology, called "context ontology". We will first explain the structure of a context attribute, as we constructed it in the definitions and then clarify how we used that structure in building the context ontology.

Context ontology
What we propose is to define a concept called "context attribute", which will encapsulate all needed information for defining and instantiating a context attribute (corresponding to our definition of a context attribute in section "Basic definitions"). Instead of treating each attribute differently (based on its complexity), we define a class (called #ContextAttribute) that will typify a common description for all context attributes: the name of the attribute, the type of needed parameters(entities) for the instantiation, the V_a (values domain), if the attribute is allowed to take a number of values (or just one) (see Table 1 for the detailed description of the class). Each context attribute will be characterized by these properties, with different restrictions: the attribute "*RoleOfPersonInGroup*" will need a #Person and a #Group as entities (parameters) and will return a #Role when instantiated, while "*NamePerson*" will need #Person as a parameter and will take as value a String, and so on.

Table 1. Description of properties of the #ContextAttribute class

Property Name	Property Type	Domain	Range	Multiple values
name	Datatype	#ContextAttribute	String	No
noEntities	Datatype	#ContextAttribute	Integer	No
entitiesList	Object	#ContextAttribute	#Entity	Yes
valueType	Object	#ContextAttribute	#Entity	No
multipleValue	DataType	#ContextAttribute	Boolean	No

In Table 1. we give the description of the #ContextAttribute class as defined in MySAM ontology. As it can be seen, the class is a general description of the characteristics of all context attributes. Each attribute is afterwards represented as an instance of #ContextAttribute, with restrictions on each property, in order to have a clear description of the specified attribute. There is some redundancy between noEntities and entitiesList, as noEntities is the cardinal of the before mentioned list. We preferred defining it explicitly as being an important property of a context attribute, even if it could be deduced by counting the elements of entitiesList.

For instance, the context attribute *RoleOfPersonInGroup* is described with the following restrictions on the properties of the defined class #ContextAttribute: Name = "*RoleOfPersonInGroup*"; NoEntities = 2 (we need to connect this attribute to one #Person and one #Group); valueType = #Role (value for this attribute is an instance of the class #Role); multipleValues = "false" (a person can only play one role in a group); entitiesList = { #Person; #Group} (connected entities - of type #Entity - are instances of class #Person and of class #Group). Here is the OWL description of the context attribute characterizing the role played by a person in a group.

```
<ContextAttribute rdf:ID="roleOfPersonInGroup">
        <nameAttribute>rolePlayed</nameAttribute>
        <noEntities> 2 </noEntities>
        <entitiesList rdf:resource="#Group"/>
        <entitiesList rdf:resource="#Person"/>
        <valueType rdf:resource="#Role"/>
        <multipleValues>false</multipleValues>
</ContextAttribute>
```

Given this structure, we can see the *valueOf* of a context attribute (see section "Basic definitions") like a function that takes as parameters *noEntities* entities (where each entity has the corresponding type defined in *entitiesList*) and returns one (or several) values of type *valueType*. Therefore, we can express a context attribute in the following manner:

$$\text{nameAttribute (entitiesList)} \rightarrow \text{valueType}(^*)$$

meaning that the valueOf function of the context attribute "*nameAttribute*" needs as parameters a list of entities corresponding to the types mentioned in *entitiesList* and returns one (or several – this is what the star *(*)* stands for) values of type *valueType*.

In Table 2. we present a small part of the list of context attributes defined in the OWL ontology used for MySAM application. We use as notation the formula described above.

Table 2. Some of the context attributes defined in MySAM ontology

Person – related	Time-related
InterestsPerson :(Person)-> String	TimeZone : (Time) -> Integer
IsSupervisorOf :(Person, Person)-> Boolean	DayOfWeek : (Date) -> String
StatusPerson :(Person) -> String	TimeOfDay : (Time) -> String
Supervises : (Person) -> Person*	
RoleOfPersonInGroup :(Person,Group)-> Role	
Location - related	**Activity – related**
PersonIsInRoom : (Person, Room) -> Boolean	ActivityStartsAt:(Activity)->Time
PersonIsAtFloor : (Person, Floor) -> Boolean	ActivityEndsAt :(Activity)-> Time
PersonIsInBuilding:(Person,Building)-> Boolean	AcivityGoal : (Activity) ->String
PersonIsInCity : (Person, City) -> Boolean	ActivityDuration :(Activity) -> Integer
	ActivityParticipants: (Activity) -> Person*
Agenda - related	**Environment – related**
BusyMorning : (Agenda) -> Boolean	DevicesAvailableInBuilding : (Building) -> Device*
BusyAfternoon : (Agenda) -> Boolean	DevicesAvailableInRoom:(Room)->Device*
BusyEvening : (Agenda) -> Boolean	DevicesAvailableAtFloor : (Floor) -> Device*

The attributes are grouped based on their specificity: attributes that are connected to a person, attributes related to time management, etc. Each attribute is described by its name and then the list of parameters needed for its instantiation (given by their type) and the type of value it will have when instantiated. A star (*) near to the type of the value means that the attribute is allowed to have multiple values for an instantiation (for example, one person can belong to several groups at a time).

This way, our ontology is made of two components: one that defines the domain (similar to all other "context" ontologies which are, in fact, *domain* ontologies) and another one, which is the description of all context attribute that will be managed in the current system (by the available Context Manager) – which is the actual *context* ontology in our case.

The representation extension that we propose allows a homogeneous representation of all context attributes in a system, therefore facilitating a generic management of those attributes, and not an ontology-dependent one. This representation has an increased expressiveness as it allows for any type of context attribute (no matter how complex) to be represented. So far, we do not perform any ontological reasoning on the ontologies that we defined, but we used ontologies (and not just XML) to define context attributes in the idea that this type of reasoning will be included in our approach. The part where this kind of reasoning will be useful is when we need to find associations between concepts defined in different context ontologies. A translation between context ontologies would be very helpful taking into account that it is highly plausible that two systems will not use exactly the same ontological description for the same context or domain.

4   Context-Aware MAS Architecture

The proposed layered architecture is composed of mySAM agents (fig. 4), each of them assisting one user able to interact with local CMs (fig. 3), each of the CMs managing one domain ontology and a set of context attributes. Being connected to the current state of the environment, each CM provides agents with context related information available in the system where mySAM is located at that time. The CM (and not agents) is the one that computes the values of all context attributes in the environment. Agents learn how to recognize relevant context and how to act accordingly. We underline the fact that this architecture is suitable for developing context-aware agents that are light-weighted, therefore adapted to hand held devices. This is due to the division of functionalities: context *management* done by an external entity (not the agent itself) and all task-related *reasoning* and *learning* done at the agent level.

We describe now the CM and the context-based agents that learn how to choose and how to use context.

4.1   Context Manager (CM)

The main functionalities of a CM (fig. 3) are: to let the agents know which is the context attributes set (defined in the ontology) that it manages, and to compute the corresponding *IRAS* from the *RAS* given by the agents at some point of processing. When entering a society, an agent asks the corresponding CM to provide it with the context attributes that it manages. Acting as intermediary between agents and the environment, CM is able to answer requests regarding its managed context attributes.

The Context Knowledge Base contains the ontology of the domain, defined as a hierarchy with #Entity as root, on one hand, and all context attributes (defined as sub classes of the class #ContextAttribute) that will be managed by the CM, on the other hand. The *instantiation* module computes the *IRAS*(f) for a given *RAS*(f). The *dependencies* module computes the values for derived attributes by considering possible relations between context attributes. For example, it can instantiate the "IsSupervisorOf" context attribute, by using the value of "Supervises", as follows:

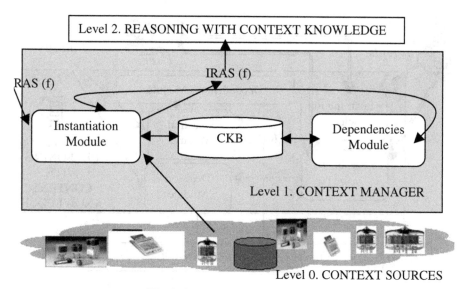

Fig. 3. Context manager architecture

IsSupervisorOf (Person1, Person2) = *true*, iif Person2 ∈ Supervises(Person1). This means that a person Person1 is considered to be the supervisor of another person Person2, if and only if Person2 is included in the set of persons supervised by Person1. In this case, we call "Supervises" a *direct* (or *primary*) context attribute, and "IsSupervisorOf" a *derived* context attribute, as it needs the value of another direct (or derived) context attribute to be instantiated.

4.2 Context-Aware Learning Agent (mySAM)

The context-based agent architecture that is the core of a MySAM agent is general and it is not restrained to the kind of application considered to illustrate our approach. Even though the agent has several other modules (negotiation, interaction, user interface, etc.), in what follows we will focus on the *selection* and *decision* modules(fig. 4), as these are the ones dealing with contextual information. The *selection* module has as functionality the selection of relevant attributes for a certain finality f (**RAS**(f)). It is the goal of this paper to show that making this module explicit and adapt its behavior dynamically can improve the system's performances.

For example, for a finality relative to deciding whether accepting or not a "2 participants" meeting, the *selection* module builds the **RAS**={"ActivityScheduled InSlot", "roleOfPersonInGroup"}; or, for a finality relative to a "seminar", the **RAS** can be {"ActivityParticipants", "Activity Description", "PersonInterests", etc}.

The *decision* module decides whether to accept or reject a meeting based on the **IRAS** resulted from the respective **RAS** instantiation. For example, if we have nothing planned for that time slot and if the person that proposes this meeting is our hierarchical superior, the decision module will propose to accept the meeting.

Both the selection and the decision modules have associated learning methods (individual as well as cooperative learning) that improve their behavior. We use the term "individual" learning to designate the learning done by the agent only in relation

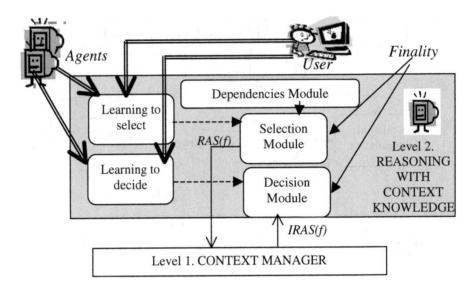

Fig. 4. Context-aware agent architecture

with its user. "Cooperative" or "multi-agent" learning is realized through exchanging knowledge (concerning either the *selection* task or the *decision* one) between agents. Using individual learning and knowledge sharing techniques, agents are able to make a better selection of **RAS** sets for a given finality and to take better decisions based on obtained **IRAS**.

Several approaches have been proposed recently ([31], [33]) concerning multi-agent learning. The use of individual learning methods in MAS is already known as indispensable if we desire to have adaptable and efficient agents. Multi-agent learning is being considered for quite a while already, and different methods were already proposed. What we propose in this article are methods to collectively learn to choose the "right" context and what to do with it. We argue that adding an explicit selection step (improved by learning) to the classical decision-making methods adds to the system's adaptability and performance.

5   Learning to Make Context-Aware Decisions

In what follows, we present the learning modules for the selection of relevant attributes (section 5.1) and for using context knowledge for decision-making (section 5.2). We want to underline that the presentation of these modules is made from a general point of view, without going into details in what concerns the specific learning method that is used for mono-agent learning (this choice is usually made taking into account the application). Our objective is to argue the necessity to use mono and/or multi-agent learning (as described in [26], [33]) when dealing with context(for both selection of relevant context and decision-making task) and to imagine multi-agent learning mechanisms suitable for context-aware applications. We are not proposing a new learning method, but show how using existing methods can improve the decision making process in a context-aware system.

5.1 Learning to Select What Is Relevant for a Decision (RAS)

Learning how to choose among context attributes which are the relevant ones could be very important in an application where the amount of available context information is too large and the effort needed to compute the values for all those attributes is not acceptable in term of efficiency. We improve the agent's capacity to choose the relevant information to assess this phase by an individual learning method (in connection to user's feedback) and also by a collective one.

For *individual learning* (mono-agent learning) on how to choose among context attributes those who are relevant for a given situation, the agent uses the feedback from the user. For mySAM, the user chooses attributes, which he considers as being relevant for the situation, before making a decision. When the user chooses more attributes as being relevant for a type of meeting, the agent memorizes them. Next time the agent will have to deal with the same type of meeting, he will be able to propose to its user all previously known relevant attributes, so that the user adds or deletes attributes or uses them such as they are.

For example, the agent received a proposal for a "family meeting". In the case of MySAM, the type of the proposed meeting is also the finality the agent will consider when choosing the **RAS**. Therefore, in this case, mySAM will know that to take a contextualized decision, it needs to find the **RAS**("family-meeting"), then ask for the CM to instantiate it, in order to obtain the **IRAS**("family-meeting"), and to make the decision based on this **IRAS**. **Let us** suppose that, for now, the **RAS**("family-meeting")={"ActivityStartsAt"}. MySAM will propose the user this **RAS** and ask if it is enough to make the decision based only on this set of relevant ontext attributes. The user can further add to the **RAS** the attribute "ActivityImportance", so that not only the start time of the meeting will be relevant. The new **RAS**("family-meeting") is now {"ActivityStartsAt", "ActivityImportance"}. Next time a family meeting proposal will be received, MySAM will list this new **RAS**, so that the user can choose to remove one of the attributes or to add new ones.

An improvement for the method used in individual learning (of selecting relevant context attributes) consists in using the attributes that other agents in the system have already learnt as relevant in that situation, therefore to make a knowledge sharing between agents (multi-agent learning). If an agent does not know which attributes could be relevant for the situation in which he is (probably for the first time), it can ask other agents which are the attributes which they already know as being relevant. In the same way, if an agent had not had a lot of feedback on attributes in a specific situation, it can again try to improve its set of relevant attributes, by asking for others' opinion. In the example given above, MySAM can choose to ask other agents for their **RAS**("family-meeting"), then make the union of all **RAS** received as a response to its quest, and propose this enlarged set to its user for further modifications. This method adds to the ability of mySAM to propose new possible relevant attributes. The user's task becomes easier, in the way that he just has to check for the relevance of the proposed attributes, without going through the whole list of available context attributes to select possible interesting ones.

5.2 Learning to Use Instantiated Context (IRAS) to Make Decisions

Using the **IRAS** retrieved from the CM is similar to a simple decision making process. To improve it, we used mono and multi agent learning to enhance the agent with the ability to evolve and to become more and more user-personalized.

Mono agent learning of how to use relevant context can be, therefore, any machine learning method developed in Artificial Intelligence (AI) which is suitable to the type of application that we want to develop. For mySAM we chose the Classification Based on Association (CBA) tool developed in the Data Mining II suite ([6]). In the "Implementation and Results" section we explain our choice and we give further details in what concerns this method.

The difficulty of multi-agent learning is due to the fact that agents have to have a common apprehension of the manner in which to use context attributes and knowledge. To be able to understand what other agents say, every agent must use the same 'language', the same manner of encoding information. They will therefore need to share at least some parts of the domain ontology, to make sure that agents understand the contents of exchanged messages.

For multi-agent learning on how to use **IRAS**(f), we chose a very similar knowledge sharing method to that presented before. The protocol proposed for this distribution can be the same, the difference consists in the knowledge that we choose to share with others: we can choose to share only the solution to the problem and not reveal the way we reasoned to find that solution, or we can share directly our problem-solving method, so others can use it for themselves. The choice depends on the system. If agents should be cooperative and it is considered that no privacy matters should be taken into account, then the second solution could be faster, in the sense that next time the agent can simply use the rule, and not ask again for the solution. But if, as considered in mySAM, the agents should not share all their criteria for accepting or rejecting a meeting (for sure we don't want the boss' agent to know that we refuse the meeting because we don't feel like getting up that early...), then sharing just the solution (in mySAM, an "accept/reject" decision) could be preferable.

Using both individual and multi-agent learning for choosing and using relevant context in making decision improves our system's openness. Agents can come in and out of a society of agents with the context adjusting accordingly and without causing problems to the general use of the system. The decision making process is as well an individual process as a collective one. Agents try to solve the agenda managing problems on their own first, and then ask for opinions and propose a solution accepted by a majority of agents in the system, based on a voting procedure.

We underline the fact that, for choosing **RAS**(f), as well for making decisions based on the **IRAS**(f), the user has the possibility to approve or reject the agent's proposal. What MySAM does is to propose solutions (actions), but it is always the user who makes the final decision. Of course, the goal would be to have an auto-customizing agent that will finally get the user's trust and will act on its own. In decision support system the idea of autonomy is not to be considered; the system is supposed to support the decision making process, not to bypass the user's approval.

However, for multi-agent systems, one of the challenges is to develop autonomous agents that are able to accomplish tasks as a replacement for human involvement. For now, we propose a decision *support* agent (not an *autonomous* one), considering that giving the user the possibility to reject the agent's proposals helps increasing the chances for the agent to be accepted without difficulty. In what follows, we detail implementation issues and give some results we obtained when testing MySAM application.

6 Implementation and Results

In order to validate our proposal, we developed a case study system (as detailed in section 2), using a multi-agent system containing several mySAM agents and one CM. To deploy our agents we used the JADE platform ([11]). JADE (Java Agent DEvelopment Framework) is a software framework fully implemented in Java language. The agent platform can be distributed across machines (which not even need to share the same operating system) and the configuration can be controlled via a remote GUI. Besides being the most used platform in MAS community at this time, JADE also proposes an add-on for deploying agents on hand-held devices, called LEAP.

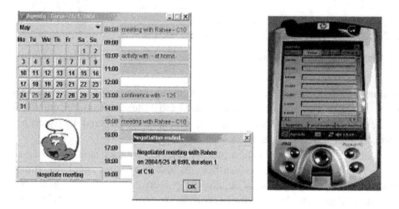

Fig. 5. MySAM interface – on PC *(left-hand side image)* and on HP Pocket PC *(right-hand ide image)*

We used JADE to develop mySAM agents, capable of handling the negotiation task for scheduling meetings. Each mySAM agent is a JADE agent with a graphical interface (see Fig. 5) that allows a user to manage her agenda. This graphical interface has been simplified for mySAM agents on a HP iPAQ 5550 Pocket PC. MySAM agents deployed on PDAs execute only the negotiation task, without any learning methods. Learning is done by a remote agent situated on a PC. This makes multi-agent sharing process even more crucial, as an agent might find itself disconnected from its remote agent that deals with all learning and reasoning modules. In this case, the agent is forced to ask for help from other agents, in order to be able to respond to its user's needs.

As soon as an agent receives a meeting proposal, it lists to the user the context of that proposal (see Fig. 6), which contains: the proposal, the known relevant attributes to that situation (**RAS**(f)), their instantiation (**IRAS**(f)) and the suggested decision (accept/refuse). The suggestion is written in red in the interface – circled on the print

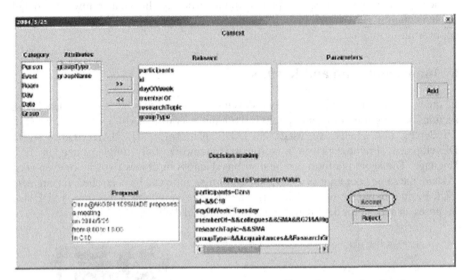

Fig. 6. Interface for selecting relevant context attributes for a precise situation

screen image in Fig. 6 (in that case, for example, mySAM proposes to accept the meeting). The user can add or remove context attributes to/from current **RAS**(f) and restart the instantiation process (to obtain the new **IRAS**) and the search for the appropriated decision in that case.

Each meeting proposal that is accepted or rejected by user is added to an example database. This example base is further on used by the agent to improve its performance (in predicting user's decision) by learning. MySAM agents use CBA (Classification Based on Association) algorithm to *learn how to use* relevant context (for acceptance or refusal of meeting proposals).

CBA (v2.1) has many features, the one that interests us being the classification and prediction using association rules. It builds accurate classifiers from relational data, where each record is described with a fixed number of attributes. This type of data is what traditional classification techniques use, e.g., decision tree, neural networks, and many others. It is proved (see [6] for details) to provide better classification accuracy (compared to CBA v1.0, C4.5, RIPPER, Naive Bayes). We used this method also because it generates behavior rules that are easily understood both by agents and humans. We have developed a module for the conversion of the rule from the CBA format into a CLIPS format. CBA format is as follows:

```
Rule 6:    busyAfternoon = false
           durationEvent = 60
           groupType = work
                -> class = yes
```

CLIPS (see [3] for details) is a productive development and delivery expert system tool which provides a complete environment for the construction of rule and/or object based expert systems. Although CLIPS also provides two other programming paradigms: object-oriented and procedural, we used the simple rule based one. Rule-based programming allows knowledge to be represented as heuristics, or "rules of thumb," which specify a set of actions to be performed for a given situation. A rule in CLIPS is specified in the following style:

```
(defrule Rule6
     (busyAfternoon  false)
     (durationEvent  60)
     (groupType  work)
               => (store CLASS yes))
```

"Class" specifies whether the agent should accept or refuse the proposed meeting. After transforming rules obtained with CBA in CLIPS rules, we could use Jess ([13]) inference engine. Using Jess, one can build Java software that has the capacity to "reason" using knowledge supplied in the form of declarative rules. Jess is small, light, and one of the fastest rule engines available.

Each agent has a module that deals with the rules generated by CBA in order to find out which rule can be applied in a specific context. The process is the following: giving a finality f, mySAM chooses the $RAS(f)$, then asks for the $IRAS$ associated with the RAS (the request is made to the CM).

The obtained $IRAS$ constitutes the current relevant context to be taken into account when making decisions. MySAM asserts as facts the current context in the form:

```
(defrule startup  =>
     (assert ( "attribute_1"   "value_attribute_1"))
 ...  (assert ( "attribute_n"   "value_attribute_n"))  )
```

Then, mySAM starts Jess inference engine to find the value of "CLASS". The value can be 'Yes' (meaning that the meeting proposal should be accepted), 'No' (meaning that the meeting proposal should be refused) or 'Unknown' (no rule matched the specific context).

When no rule matches the specific context (no rule is activated when using Jess), mySAM uses *a multi-agent knowledge-sharing* process to find out how to use this specific context ($IRAS$) to decide. The agent starts a voting system: it asks all known agents in the system for their opinion on the situation (the situation being defined as $\{f,IRAS(f)\}=(f,\{(attribute\_1, value\_attribute\_1),...,(attribute\_n, value\_attribute\_n)\})$) and counts each opinion as a vote for "accept", "reject" or "unknown". The agent then proposes to the user the decision that has the most votes. Agents consider an "unknown" result as a "reject" (by default, an agent will propose the user to reject all meeting proposals that neither it, nor other agents know how to handle). We chose to use this "voting" procedure because it was easier than trying to share the rules for themselves and because, in this kind of application, the privacy is important. Not all agents will want to share their decision-making rules, but an "accept/reject/unknown" answer is reasonable to be shared. In this way, agents do not explain how they inferred that decision, but just what the decision is. However, if rule sharing is to be taken into account, there are several important details that should be addressed: using common vocabulary, understanding the encoding of the rule, managing privacy, etc.

The context manager (CM) is also implemented as a JADE agent. CM is a special agent in the system that has access to the domain ontology that defines the concepts that characterize the domain and the context attributes that the CM will manage. CM provides context knowledge to all agents that are active in the system. The ontology was created using Protégé 2000 (Protégé is a free, open source ontology editor and knowledge-base framework [22]) and the agent accesses the ontology using Jena ([12], a Java framework for building Semantic Web applications.

Agents interactions in the system are quite simple: mySAM agents can query the CM using a REQUEST/INFORM protocol, the meeting negotiations between mySAM agents are done in a PROPOSE/ACCEPT/REJECT manner.

When testing mySAM we were able to draw several conclusions, such as:

- using a selection step to choose the **RAS** for a situation helps in having smaller and more significant rules. Using all attributes to describe a situation is not only difficult to deal with, but also unnecessary. We tested our hypothesis on a set of 100 examples. For 15 context attributes used, we obtained an overall classification error of 29.11% and more than 40 rules. When we split the example set on several finalities ("family-meeting", "friends-meeting", "work-meeting"), the user choses for each situation a limited number of context attributes (an average of 7 attributes for a meeting with family, 11 for the two others); the error, tested on the same set of examples, with the same classifier, but taking into account only those *relevant* attributes selected by the user, becomes 7.59% and the number of obtained rules drops to an average of 15;

- we consider that sharing with other agents just the solution (accept/reject) for a situation is enough, because the agent that received the answer will then add this situation to its examples base, from where it will then learn the appropriate rule. Even if it will take longer to learn that rule than just having it immediately provided by others, the privacy problem is this way solved, because we share just the answer to a specific situation, and not the reasoning that produces such an answer. Also, the user has the possibility to decide otherwise (if the decision is not suitable for him/her). Whereas, if agents shared rules and they added those rules directly into their reasoning modules (without user's consent), this could introduce a significant amount of errors in agent's decision making.

7 Related Work

Our study is centered on multi-agent systems. However, our definition and vision on context is based more on approaches in other domains, like Artificial Intelligence, decision support systems, human-machine interaction, ubiquitous computing, etc.

Definitions. Persson [21] sees context as "[...] the surrounding of a device and the history of its parameters", Brezillon[1] considers it as being "the objective characteristics describing a situation,[...] the mental image generated, [...] the risk attitude" used to make decisions and Turner [30] defines it as "any identifiable configuration of environmental, mission-related and agent-related features" used to predict a behavior. We proposed in section 2.1. some definitions that are quite similar to all these in the sense that are based on: (i) the elements that compose the context and (ii) its use, i.e. the finality (the rationale we pursue) when using this context. The definition we proposed takes into account those two dimensions of context (its use and its elements) and it explains what each dimension is and how to properly define it when designing a context-based MAS.

Models and architectures. Concerning existing models and architectures of context-aware applications, there are several models that were proposed recently. We detail here some of them: Gaia ([Ranganathan and Campbell [23]]), CoBrA ([Chen et al. [2]]) and Socam ([Gu et al. [28]]). Gaia proposes an infrastructure that allows applications to obtain contextual information and to reason over context. The infrastructure contains: context providers – which obtain context from sensors or data sources; context consumers (context-sensitive applications) that query the providers for information; context synthesizers get sensed contexts from providers and derive higher level (abstract) contexts and provide them to applications. Some context history is stored in a database.

CoBrA ([Chen et al. [2]]) is centered on a context *broker* that acquires contextual information from heterogeneous sources. This information is afterwards assembled in a coherent model and the context broker will share this model with all computational entities located in that same environment.

The notion of "context provider" is again used in SOCAM ([Gu et al. [28]]), to define entities, which will abstract contexts from external or internal sources and will convert them into an OWL representation. In this way, context may be shared between other architectural components. The central entity is this time a context *interpreter*, composed of a context reasoning engine and a context knowledge base (KB) which inspired our Context Manager architecture. The context reasoner has the functionality of providing deduced contexts based on direct contexts, resolving context conflicts and maintaining the consistency of the context KB. The Context KB provides a set of API's for other service components to query, add, delete or modify context knowledge. The Context KB contains: context ontologies in a sub-domain and their instances.

Meeting schedulers and context. In MAS, the notion of context is used to describe the factors that influence a certain decision. In similar applications (meeting schedulers), context means: type of event, number of attendees, etc. (Calendar Apprentice [18]), activity, participants, location, required resources (Personal Calendar Agent [15]), system load, organization size (Distributed Meeting Scheduler [26]), time, user's location, etc. (Electric elves [25]). None of these works mention the idea of context but they all use "circumstances" or "environmental factors" that affect the decision-making.

In the Calendar Agent ([14]), Lashkari *et al.* use the notion of context, but they assume that relevant context is known in advance, so that all contextual element that they have access to is considered relevant for the decision to be made. This is the major problem with the way context is used: these approaches are not fitted for an application independent way of handling context, because they do not provide a general representation of context knowledge and methods to choose between relevant and non-relevant context elements for a specific decision.

8 Discussion

The main difference and contribution of our work is in the sense that we propose a MAS architecture based on an ontological representation of context and that can permit individual and multi-agent learning on how to choose and how to use context. It is not the choice of an application that generated this architecture, but MySAM is just a case study to test our approach. We can see that the architecture and model we proposed could be very well used for similar applications: applications that need user-centered, personalized, adaptable behaviors.

All that changes with the application scenario itself is the domain and context ontology. As long as the application still needs some context management mechanism, the designer only needs to modify the ontologies so they reflect the current domain knowledge (that could be medicine, mechanics, environmental science or whatever) and the context information that is available in the system (temperature, blood pressure, tools, maps, etc.). Once this is done, the context manager is able to manage the newly defined context without problems. Agents are still able to reason on how to choose what is relevant for their task, no matter what the task actually is. They are still using other agents' experience to improve their own and learning methods to better adapt to their user and their environment.

Nonetheless, some issues still need to be considered when using our approach. Forcing the user to deal with a (large) list of possibly relevant attributes is not realistic; we tried to ease user's task by presenting attributes regrouped by categories, but further improvements must be done.

Sharing private information with others imposes privacy problems: agents are supposed to share information about their user's criteria when accepting or rejecting a meeting. This sharing session might not be welcomed by the user, even if it is highly motivated by the increased speed of the agent's learning. The privacy issues must be dealt with, at least by letting the user know when and what information the agent is about to share and with whom.

A big amount of communication overload might be generated, as all agents ask for help from all other agents. If the system will consist of several dozens (or hundreds) agents, then the communication overload would be considerable. A solution can be for each agent to find 'specialists', a specialist being an agent that has the most knowledge about how to solve a particular situation - and use only those agents in case of need. (see P.Maes for a very similar idea in [16]).

Asking for user's feedback for each step of the way could be a burden for the user. This can be seen as a drawback, but also as a good thing, as in this way the user keeps the control of the application. We can imagine that after a while, the user will be much more comfortable delegating tasks to the agent if she had observed that the agent really behaves according to her needs.

Another issue to be considered is the information that might be missing: the user might want to consider some information as relevant, but not be able to find it in the list of available attributes. We think that the agent should give the opportunity to the user to add specific attributes that she might find relevant, and, furthermore, be able to translate and compute this information when moving to another system. For this, an ontology-based correspondence between different attributes of context must be done by the CMs. All these issues will need to be considered in future work.

9 Conclusions and Future Work

In this paper, we have presented a definition of the notion of context, notion that is used in most of the systems, usually without precisely and explicitly taking it into account. We have proposed an ontology-based representation for context elements and a context-based architecture for a learning multi-agent system that uses this representation. We then tested our approach by implementing a meeting scheduling MAS that uses this architecture and manages and learns context based on the definitions and representation we proposed. We obtained encouraging results by embedding the selection of relevant information into the agent itself, showing that this improves the agent's performance. Furthermore, adding just a very simple learning method to the selection module makes the agent behave more appropriately, by revising its behavior dynamically.

As future work, we aim to extend this framework for context-based MAS to be used for any kind of application that considers context when adapting its behavior. The context manager has to be able to deal *automatically* with all context-related tasks (including the computation of context attributes values) and to share all his context-related knowledge. In order to make this possible, our future work will focus on representing and managing: (i) how to compute the values for derived context attributes (how to define the *valueOf* function), (ii) dependencies that can exist between context attributes (how to compute higher level context information from lower level one), (iii) the importance of different attributes in different situations (making a more refined difference between relevant and non relevant attributes, using a degree of relevance, not just relevant/non-relevant distinction).

In what concerns learning agents, the framework will provide agents with at least one individual learning algorithm and all that is needed to communicate and share contextual knowledge (how to choose, compute and use context to make decisions).

In order to prove the generality of our approach, we will implement it to solve different other scenarios of context-aware applications that can use multi-agent technology.

References

1. Brezillon, P. – "Context Dynamic and Explanation in Contextual Graphs", In: Modeling and Using Context (CONTEXT-03), P. Blackburn, C. Ghidini, R.M. Turner and F. Giunchiglia (Eds.). LNAI 2680, Springer Verlag (http://link.springer.de/link/service/series /0558/tocs/t2680.htm). pp. 94-106.
2. Chen, H., Finin T., Anupam J. – "An Ontology for Context-Aware Pervasive Computing Environments", The Knowledge Engineering Review Volume 18 , Issue 3, p. 197 – 207, 2003.
3. CLIPS (C Language Integrated Production System) - http://www.ghg.net/clips/ CLIPS.html (last visited: 10 Oct. 2005)
4. Coutaz J., Ray G. –"Foundations for a theory of contextors", In Proc. CADUI02, ACM Publication, pp. 283-302, 2002.
5. Coutaz J. et al – "Context is key", Volume 48 , Issue 3 (March 2005) The disappearing computer, SPECIAL ISSUE: The disappearing computer, Pp: 49 – 53, 2005.

6. Data Mining II – CBA - http://www.comp.nus.edu.sg/ ~dm2/ (last visited: 10 Oct. 2005)
7. Dey, A., Abowd, G.– "Towards a better understanding of Context and Context-Awareness", GVU Technical Report GIT-GVU-00-18, GIT, 1999.
8. Edmonds B. – "Learning and exploiting context in agents", in proceedings of The First International Joint Conference on Autonomous Agents and Multi-Agent Systems, AAMAS 2002, July 15-19, 2002, Bologna, Italy, pag. 1231-1238.
9. Gonzalez A., Ahlers R. – "Context based representation of intelligent behavior in training simulations", Transactions of the Society for Computer Simulation International, Vol. 15, No. 4, p. 153-166, 1999.
10. Henricksen K., Indulska J., Rakotonirainy A.– "Modeling Context Information in Pervasive Computing Systems", Proc. First International Conference on Pervasive Computing 2002, p. 167-180, Zurich.
11. JADE (Java Agent Development framework) : http://jade.cselt.it/ (last visited: 10 Oct. 2005)
12. Jena Semantic Web Framework - http://jena.sourceforge.net/ (last visited: 10 Oct. 2005)
13. Jess: http://herzberg.ca.sandia.gov/jess/index.shtml (last visited: 10 Oct. 2005)
14. Lashkari Y., Metral M., Maes P.– "Collaborative Interface Agents", Proc. of the Third International Conference on Information and Knowledge Management CIKM'94, ACM Press, 1994.
15. Lin S., J.Y.Hsu – "Learning User's Scheduling Criteria in a Personal Calendar Agent", Proc. of TAAI2000, Taipei.
16. Maes P – "Agents that reduce work and information overload", Communications of the ACM, Vol 37, No 7, July 94.
17. Matsatsinis N.F., Moratis P., Psomatakis V., Spanoudakis N. – "An Intelligent Software Agent Framework for Decision Support Systems Development", ESIT ´99 (European Symposium on Intelligent Techniques).
18. Mitchell T., Caruana R., Freitag D., McDermott J., Zabowski D.– "Experience with a learning personal assistant", Communications of the ACM, 1994.
19. OWL - http://www.w3.org/2004/OWL/ (last visited: 10 Oct. 2005)
20. Ossowski S., Fernandez A., Serrano J.M., Hernandez J.Z., Garcia-Serrano A.M., Perez-de-la-Cruz J.L., Belmonte M.V., Maseda J.M. – "Designing Multiagent Decision Support System. The Case of Transportation Management", ACM/AAMAS 2004, New York, pp 1470-1471.
21. Persson P.– "Social Ubiquitous computing", Position paper to the workshop on 'Building the Ubiquitous Computing User Experience' at ACM/SIGCHI'01, Seattle.
22. Protégé 2000 - http://protege.stanford.edu/ (last visited: 10 Oct. 2005)
23. A. Ranganathan and R. Campbell - "An infrastructure for context-awareness based on first order logic", Personal and Ubiquitous Computing, 7(6):353--364, 2003.
24. Ryan N.– "ConteXtML: Exchanging contextual information between a Mobile Client and the FieldNote Server", http://www.cs.kent.ac.uk/projects/mobicomp/fnc/ConteXtML.html (last visited: 10 Oct. 2005)
25. Scerri, P., Pynadath D., Tambe M.– "Why the elf acted autonomously: Towards a theory of adjustable autonomy " , First Autonomous Agents and Multi-agent Systems Conference (AAMAS02), p. 857-964, 2002.
26. Sen S., E.H. Durfee – "On the design of an adaptive meeting scheduler", in Proc. of the Tenth IEEE Conference on AI Applications, p. 40-46, 1994.
27. Sian S. S. – "Adaptation Based on Cooperative Learning in Multi-Agent Systems", Descentralized AI, Yves Demazeau & J.P. Muller, p. 257-272, 1991.

28. Tao Gu, Xiao Hang W., Hung K.P., Da Quing Z.– "An Ontology-based Context Model in Intelligent Environments", Proc. of Communication Networks and Distributed Systems Modeling and Simulation Conf., 2004.

29. Turney,P. – "The identification of Context-Sensitive Features: A Formal Definition of context for Concept Learning", 13th International Conference on Machine Learning (ICML96), Workshop on Learning in Context-Sensitive Domains, p. 53-59.

30. Turner, R. – "Context-Mediated Behaviour for Intelligent Agents", International Journal of Human-Computer Studies, vol. 48 no.3, March 1998, p. 307-330.

31. Sian S. S. – "Adaptation Based on Cooperative Learning in Multi-Agent Systems", Descentralized AI, Yves Demazeau & J.P. Muller, p. 257-272, 1991.

32. Willmott S., Calisti M., Rollon E. –"Challenges in Large-Scale Open Agent Mediated Economie", Proc. of Workshop on Agent Mediated Electronic Commerce AAMAS 02, july 2002.

33. Weiss G., Dillenbourg P.– "What is "multi" in multi-agent learning?", P. Dillenbourg (Ed) Collaborative-learning: Cognitive, and computational approaches, p. 64-80, 1999.

A Layered Model for User Context Management with Controlled Aging and Imperfection Handling

Andreas Schmidt

FZI Research Center for Information Technologies, Karlsruhe, Germany
Andreas.Schmidt@fzi.de

Abstract. Current research in context-awareness is biased toward low-level context information. High-level context information, however, poses several challenges to context management systems, which can be traced back to the asynchronicity of context acquisition and use and the inherent dynamics and imperfection in that process. This paper presents a three layer model allowing for dealing with the problems of imperfection and aging in a controlled way. It conceives the problem of high-level user context management as an information management problem with specific requirements. The approach has been applied to a context-aware learning environment for corporate learning support.

1 Introduction

1.1 The Need for High-Level Context Information

With the ever-increasing volume of accessible information and the advent of ubiquitous information access via mobile and wearable devices, the focus of information systems research has shifted to increasing the efficiency of information access for the user. This encompasses both simplifying the query formulation process and improving the relevance of query results. In general, there is a trade-off between preciseness (and thus selectivity) of queries and ease of query formulation. The most promising approach is to incorporate information about the user and her current situation (i.e., her "context" [1]), which is the fundamental approach in context-aware and situation-aware systems.

The upsurge of interest in context-awareness has mainly occurred in the area of *low-level context information* as defined by [2], which represents information about the user's context that can be directly sensed (or obtained) or aggregated from this sensor data in a relatively straight forward way (although the concrete implementation may still pose severe problems). The most prominent examples here are location-aware systems. But recently, it has been discovered that the consideration of context is a key enabler for next-generation information services on a much broader scale. One example are e-learning and knowledge management systems ([3], [4]), which require rather high-level context information like tasks, business process steps, or the information whether a person currently is under time pressure or has some time to learn.

Apart from determining how the context influences the information need of a user, the key problem in these systems is how to keep the user context information up to date, which includes both the acquisition and the management of such information over time. In general, only indirect methods can be used for high-level context information,

T.R. Roth-Berghofer, S. Schulz, and D.B. Leake (Eds.): MRC 2005, LNAI 3946, pp. 86–100, 2006.
© Springer-Verlag Berlin Heidelberg 2006

and several sources of context information have to be considered. Some information can be derived from the user's interaction with specific applications, other pieces can be obtained from data stored in other systems, e.g., in a corporate environment from Human Resources data, Workflow Management Systems [5], or Personal Information Managers (addresses, calendar etc.) [6], still others can be retrieved on demand with specialized operations (e.g., localization in wireless networks [7]). It has been realized (see also [8]) that this complexity should be hidden from a context-aware application by establishing a "user context management" middleware (or "broker architecture" [9]) that provides applications with an up-to-date view of the user's current context, partially materializing it, partially retrieving it dynamically from other sources.

1.2   Challenges for High-Level User Context Information Management

However, the idea of a context middleware for high-level context information faces several challenges, which cannot be adequately met by existing information management solutions [10]:

- **Dynamics.** The context of a user continuously changes. Different features of the context change at different pace; e.g., name and occupation change less frequently than location or current task. Additionally, we have to distinguish between context evolution and context switching. In the latter case, part of the context changes, but it is quite likely that it later changes back again so that we should store the information for later use. Typical examples are private and professional information, project-specific context and role-specific context.
- **Imperfection.** User context information is typically collected via indirect methods that observe a user's behavior and infer facts from these observations. These methods do not yield certain results, in some case they are more, in other cases less probable (uncertainty). Additionally, some of the information cannot be determined exactly (imprecision). The most typical example here is location information. Depending on the method (GSM, GPS, etc.), the precision of the information can vary a lot, which is particularly important for the most prominent examples of context-aware services: location-based services. Additional aspects of imperfection in the area of user context information are incompleteness and (as we collect it from several sources and/or with a variety of methods) inconsistency.

Similar observations – although from a different perspective – have also been made by [11].

1.3   Overview

In the remaining part of this paper, a context information architecture is presented that is able to deal with these two challenges. First, the specific requirements posed by high-level context information to the respective management infrastructure are briefly discussed. Then a basic layered context information architecture is introduced that deals with the problems of aging and imperfection. In the following section, an extension with the concept of subcontexts is introduced that improves the handling of the dynamic nature of user context information. In section 4, there will be an overview of the prototypical implementation in the project "Learning in Process".

The notion of *user context* as used within this paper encompasses both more stable information, which is often called the *user profile*, and more volatile information about the user's or environmental state. The distinction between the two of them is conceptually hard, and due to flexibility of the approach, different dynamical behaviour can be modelled so that there is also no need to distinguish.

1.4 Specific Requirements for a Suitable Context Model

The context information architecture proposed in this paper focuses on the so far neglected issue of high-level context information as defined by [2]. This differs from mainstream context-awareness research because it poses special requirements to a context management infrastructure. The differences can be traced back to a main distinction: *Asynchronicity*. In contrast to low-level context information, high-level context information typically cannot be continuously monitored (via sensors) or determined on demand at any instant in time (e.g. GPS positioning). Rather – due to the required complex abstraction process –, the system has to collect in advance information about the user and make it persistent over time. This "materialized" approach introduces several problems (which can be ignored in low-level context settings):

- **Aging.** As a direct consequence of the dynamics of real-world situations, it should be obvious that collected context is not valid indefinitely. If the system gets to know about the current "task" of the user, this information will only be valid for a limited amount of time. As a consequence, the user context management system needs to have some aging mechanism as a form of controlled forgetting. This controlled forgetting also limits the impact of incorrect context data, which considerably increases the robustness of the system.
- **Variability in dynamic behavior.** The closer inspection of the "aging" problem reveals that aging is not uniform across the different aspects of user context information. While information like name, birthdate changes infrequently or never, other aspects like personal skills, interests goals evolve over time, and tasks or location are highly volatile. So the aging support has to be specific for the different parts of the context. Additionally, there may also be context data which is known to be valid only for a definite time interval.
- **Reasoning over time.** Although it could drastically simplify the design of the management infrastructure, it is usually not sufficient to just keep the *current* context. Rather, many methods ranging from conflict resolution to context augmentation (e.g. through inferencing techniques) do not solely rely on the current user's context, but also on the history in order to detect patterns or to use the history as a reinforcement.
- **Scalability.** If we want to use context information for applications like corporate learning support, the context management system needs to be scalable with respect to large numbers of users and long time frames. This is especially critical as the research on artificial intelligence in general and the Semantic Web in particular has shown that there is trade-off between expressiveness and scalability of the methods.

2 Layered Context Model

2.1 General Considerations

Current approaches to context modelling like [12] or [13] and to applying context awareness to the e-learning and similar domains like [14], [15], [16] emphasize the potential of applying Semantic Web technologies to user context management. It enables the creation of more semantically aware processing methods, especially by introducing a shared vocabulary, which can be used across different tools and systems, and by applying reasoning techniques based on domain knowledge. But Semantic Web technologies have still quite a way to go for solutions that are comparable in terms of scalability with traditional information management solutions. To take that into account, the context model presented here was developed as a data model (in the tradition of relational or object-oriented databases) whose core does not depend on those reasoning techniques, but which can integrated smoothly with those techniques if a domain-specific schema requires it.

The second issue is directly related to the architecture of user context information management systems. For traditional database management systems, it has proven effective to divide the management functionality into different layers which are basically independent of the internal logic of the lower layers. In that spirit, we want to present a three layer model that allows for structuring the problem in a better way (see figure 1):

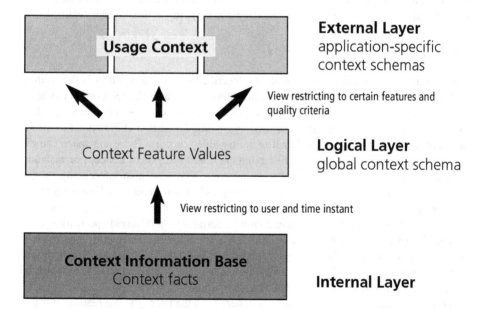

Fig. 1. Layered Context Model

- **Internal Layer.** The internal layer stores all collected information about users in a time-dependent way. It may contain contradictions and redundancy.
- **Logical Layer.** This layer provides a complete view on the context of a user at a specific instant in time together with the imperfection metadata that allows for determining the reliability of the stored information. However, inconsistencies in the collected data are removed with respect to the considered instant in time using conflict resolutions strategies.
- **External Layer.** This layer represents the usage context of a particular application at a certain instant of time. The context information is in a schema the application understands, and also certain quality criteria (like minimum confidence) are guaranteed.

This layering scheme assumes that most applications, but also the majority of context data sources will access the system through the external layer, using their own schema. They don't have to deal with the complexity of the context data and the management problems tied to it. In contrast to having each application resolve conflicts and calculate the reliability of collected information, this task can be centralized in a generic middleware for user context management. This middleware can be enhanced on the lower layers by adding appropriate heuristics and plugging in reasoning services for dealing with ontological background information.

In analogy to database systems, we begin with describing the logical layer (which corresponds, e.g., to the relational data model), then explain the internal layer and how the internal layer maps to the logical layer, before explaining the external layer (which corresponds to the concept of views) and how the mapping to this layer is realized.

2.2 Logical Layer

The primary construct of the data model to describe the context of a user is the *context feature* (cp. [17]), which corresponds to attributes (in case of the relational or object-oriented data model) or properties (in case of RDF(S) and other conceptual data models like OWL). Context features are described by a unique identifier (URI), a value space, and cardinality constraints (whether it is multi-valued or not). The value space can either consist of atomic data types (like numbers, dates etc.) or of concept or instance identifiers referencing elements from an ontology. That way, also inferencing on the value level is possible by reusing description logics based reasoners and ontology management systems.

In order to allow for better reusing context information in different applications, the model also offers the possibility to define a feature hierarchy via feature inheritance, which directly corresponds to property hierarchies in RDF(S). This adds a basic inferencing capability to the model: if an applications requests the value(s) for a specific feature, the values of sub-features can be also returned.

Feature values are tuples (U, f, v, α), where U is the user, f is the feature, v is the value and α is the confidence level that the feature f currently has the value v for the user U.

An example would be *(Andreas, performs-task, literature-search, 0.8)*, which encodes that the user *Andreas* currently performs the tasks of *literature-search* with the probability of 0.8, which could have been determined from monitoring his usage of a web browser and the visited sites. *literature-search* could reference to a concept in an ontology that allows for generalizing this concept to *search*.

The operations supported at this layer are

- *queries* for specific context feature values of a user and for users having certain feature values
- *update operations* that can set or delete feature values for a specific user

2.3 Internal Layer

In contrast to the logical layer, the internal layer does not only have to provide the current view on the user context, but it also needs to store the history so that temporal concepts need to be considered. The internal layer primarily consists of a fact base (called *context information base*) with context facts as its entries:

A *context fact* is a tupel $(U, f, v, t, valid, \alpha)$ where

- U is a user
- f is context feature
- v is a value
- t is the point in time at which the factum was added to the fact base.
- $valid$ is the validity interval for the value
- α is the minimum probability that at point of time t the feature f has the value v for user U.

As additional schema-level information, the internal layer has *aging functions* attached to each context feature, which allow for describing how the confidence in a certain value decreases over time. An aging function basically is a function $a : TIME \rightarrow [0, 1]$, which is monotonically decreasing. The aging function is applied to the time distance, and the resulting value is multiplied with the initial confidence value in order to obtain the current confidence value. These aging functions can be assigned heuristically or – preferably – based on empirical results. For the practical implementation, aging function types are defined (e.g. linear, polynomically decreasing, exponentially decreasing) with parameters. This function type approach allows for easy configuration and/or learning of aging functions, and for more efficient optimization techniques.

Taking the example from above, the context information base could contain *(Andreas, performs-task, literature-search, [2005-04-15 10:00, ∞), 2005-04-15 10:00, 0.8)*, and an entry *(Andreas, performs-task, examine-students, [2005-04-14 14:00, ∞), 2005-04-15 13:00, 0.9)*.

The operation on this layer are much more powerful as they can additionally exploit the temporal perspective.

2.4 Mapping the Internal Layer to the Logical Layer

In order to map from the internal layer to the logical layer, the following issues need to be taken care of:

- **Restrict to a specific point in time.** The set of context facts is restricted to those context facts for which the validity interval contains the requested instant of time.
- **Apply aging functions.** With the help of aging functions, the current confidence value needs to be calculated.
- **Infer additional information.** As the context facts only represent those values that were explicitly added to the fact base, we provide also a feature hierarchy on the logical layer, for which additional feature values need to be inferred.
- **Resolve inconsistencies.** Inconsistency occurs in our model if there are multiple values for a feature for which the cardinality constraints enforce a single value. There can be different strategies to resolve these inconsistencies. The most obvious is to take the value with the highest confidence, but usually the strategy also needs to take into account that facts can be reinforced by other facts (e.g. two independent methods determine the same feature value within a limited time window).

If we apply this procedure to the example, it is clear that the restriction to a specific instant in time (e.g. *2005-04-14 11:00*) still provides two possible tasks. After applying the aging function, let's suppose that the *literature-search* has confidence 0.7 and *examine-students* has confidence 0.1. This would lead to the resolution strategy to take the *literature-search* as the current feature value, because we have specified that the *performs-task* feature is only single-valued.

2.5   External Layer and Mapping from the Logical Layer

The external layer is intended to be the interface for the application, providing an application-specific view. On the one side, this consists of an application-specific context schema, on the other side the application can specify a certain quality-level, which depends on the usage strategy. This quality level is expressed as a minimum confidence level for supplied user context information. Where user context information is only used as a rough indication about the user, the confidence does not matter that much, but for applications that involve legally binding transactions, certainty about values is crucial.

In order to perform the mapping from the logical layer, the following two issues need to be taken care of:

- **Apply quality criteria.** This involves the filtering of the available feature values according to the supplied minimum confidence.
- **Perform a mapping.** In this step, the global context schema used on the logical layer is translated into an application-specific schema. In case of simple projections and renamings, this can be done within the user context management system, but for more powerful mapping features, external mapping services are the method of choice (in spirit of [18]).

3   Subcontext Switching Support

3.1   The Concept of Subcontexts

The layered context model introduced so far supports the controlled *forgetting* of outdated context information and allows for representing different levels of certainty. However, it does not help with the problem of slow adaptation to a changed context. This is due to the phenomenon that there are often dependencies between different features, i.e. groups of values often change together. A typical example are different roles (e.g. private role or business role). If a user currently has the private role, a wide range of context information can be different, e.g. his payment preferences, contact information etc. In order to cope with these dependencies, the concept of subcontexts is introduced. Subcontexts are basically groups of feature values that change together and have the following characteristics:

- Subcontexts can be nested.
- Subcontexts conform to a schema that defines (1) available features and (2) nesting relationships.
- For each subcontext schema, there is at most one sub context active at a certain instant of time. The others are present as inactive, "parallel" sub contexts yielding the same aspects about the user.

An example for a subcontext structure is given in figure 2. Here you have a user with location-dependent, role and project-dependent information. Currently the user "John Smith" is at his office and thus has broadband network access, but no loud-speakers (which could influence whether the system can deliver video or audio material). The system also knows the characteristics of other locations, but those are represented as currently inactive sub contexts. There is also role-specific information and within each role project-specific information. In the case of corporate e-learning, this information could in its majority be provided manually by a learning coordinator or human resources department.

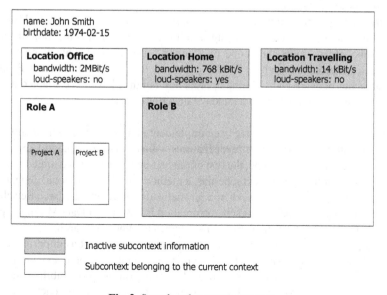

Fig. 2. Sample sub context structure

Subcontexts schemas can be seen – in contrast to functional dependencies – as weak dependencies, which can still be specified on the schema level. It does not specify the semantic key that determines the other values. A subcontext schema just states that there is some relationship. This weakness has the benefit that the schema can be learnt on the basis of context histories even under the assumption of imperfection, which will be shown in the next section. The subcontext structures can also be seen as a replacement for explicitly modeled domain knowledge that reaches beyond the current context (e.g. rules). The main benefit of the subcontext structure approach is the ability to deal with the problem of imperfection in a uniform way.

In our three layer model, the concept of subcontexts is located at the internal layer as it is an implementation technique to realize faster adaptation to changing feature values. Typical applications will not access the subcontext structure, although there could be applications exploiting the available subcontexts, e.g. providing the possibility to directly switch between them, e.g. by using methods like [19] without completely relying on that method.

3.2 Heuristic Strategies for Subcontext Management

Although it is possible and in many cases also reasonable to have the user manually indicate her current context, this limits the concept of subcontexts to a relatively coarse granularity and less frequent context switching. Ongoing research therefore tries to identify heuristic strategies how to automate the handling. The most crucial issues are:

- **Detection of subcontext changes.** The most crucial part of the research is how to detect context changes, or to be more precise: how to distinguish context evolution from context switching.
- **Automatic construction of subcontexts.** Although being quite practical for closed environment like intranet e-learning solutions, the manual specification of schemas for subcontexts limits the scope of a generic user context management service. Therefore automatic methods for sub context detection are investigated. Promising approaches here are Data Mining approaches, but they have to be adapted to (1) the scarce amount of available data and (2) uncertainty aspects.

Currently, simple strategies have been implemented. For the detection of sub context changes, the strategy works with pivot features, which serve as semantic keys for sub contexts. If these features change, the rest of the feature value is also assumed to change. For the construction of subcontext schemas, a method based on functional dependencies is used (which is borrowed from schema normalization in relational database schemas. A second strategy for subconstext schema creation is based on a Data Mining approach for discovering association rules. Further strategies are under research.

It should be obvious that detecting context switching is at least as imperfect as other methods for context acquisition. Therefore it is important that context switching leads to context facts with a reduced probability, depending on the probability that a context switch has actually occurred. Especially the association rule approach can provide (through the confidence) an indicator on the reliability.

4 Implementation

4.1 Overview

A context management system based on the model presented in this paper was implemented in the project *Learning in Process*, which aimed at – among other things – supporting a new type of learning process for workplace learning: *context-steered learning*[20]. Instead of having human resource development experts assigning courses to employees or leaving it to the employees to actively search for learning resources satisfying their knowledge need, the LIP system continuously monitors the employee's working activity (i.e. her context) and deduces from it (with the help of domain knowledge) the possible knowledge gap. Based on this gap, the system can recommend relevant learning resources, which can (but need not) initiate learning processes. The context schema used for that purpose incorporates the following features:

- **Personal characteristics:** competencies, goals, learning preferences like interactivity level and semantic density
- **Organizational aspects:** role, organizational unit, business process (step), task
- **Technical aspects:** user agent, available bandwidth, available multimedia equipment

For more details on the context schema and how it was constructed see [4].

As background knowledge, a competence catalogue was defined that specifies relevant competencies. Learning resources were conceived to have learning objectives, which were described as competencies to be acquired, and prerequisites, which were described as competencies needed to comprehend the presented resource. Furthermore, organizational aspects like roles, organizational units, process steps and tasks were annotated with competency requirements. This enables the context-aware recommendation, for which the context management service forms the basis.

4.2 Context Management Service

The basis for the service matching the current work situation and relevant learning resources is a user context management service that provides a view on the current context of the employee (see figure 2). In order to be able to easily implement feature inheritance and to smoothly work with moderately expressive ontologies used for encoding the domain knowledge, the implementation of the *internal layer* was based on the ontology management system KAON[21]. KAON supports a variation of the RDFS data model and is implemented ontop of relational databases that support (at least some variation of) SQL-99. The operations of the internal layer were exposed both via specialized methods and via a descriptive query language (based on the KAON query language). For the project prototype, however, only the specialized methods were used by external applications.

Queries to the *logical layer* are also expressed in the KAON query language – mainly because it allows for smoother interoperability with the rest of the system. The user context features are mapped to virtual properties of the RDFS data model. The queries

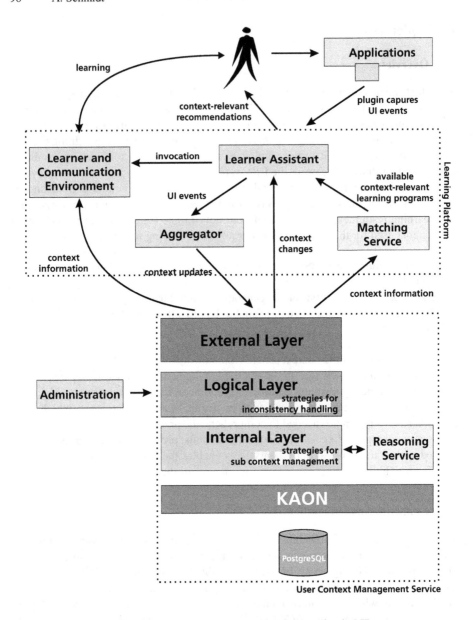

Fig. 3. Overview of the prototypical implementation in LIP

are rewritten to queries to the lower layer. Inconsistency resolution is done by postprocessing the results, but currently there is only a simple strategy in place (as described above). In the prototype, the *external layer* was very thin and did not require mapping services so far. As a consequence, there was only a confidence cut-off.

As *context sources*, on the one side there was a event-triggered update of more stable elements like name, role or organizational units. On the other side (*Administration* in figure 2), there were desktop *learning assistants* monitoring the interactions of the employee with her applications. For the two pilot installations, the following applications were observed: Internet Explorer, Microsoft Office (Excel, Word, Powerpoint), and Microsoft Visual Studio. The user interface event were aggregated and translated into context feature value changes.

In order to enable context-triggered actions, the service also supports notifications about context changes. These notifications were used by the learner assistant to determine when to invoke the *matching service* that retrieves relevant learning resources and compiles them into a personalized learning program based on the context of the user and background domain knowledge (for more details see [20] and [4]).

The simple strategies for inconsistency handling and subcontext switching were sufficient for the first prototypes. However, it is expected that with more context sources connected more sophisticated strategies will be required. The prototype provides a good basis for further experiments on this.

5   Related Work

There are plenty of models for dealing with user context information, both from the traditional community of user modeling and the recently emerged communities for context-awareness. A good overview of recent context modeling approaches gives [8]. In general, it can be stated that the data management problem is a neglected area of research.

The consideration of the imperfection and dynamics of user context information is also a relatively neglected area of research, especially for the case of high-level context information. [22] investigate quality criteria for context information complementing quality of service concepts. They define the following criteria: precision, confidence, trust level (for context sources), granularity and up-to-dateness. [23] introduce meta attributes like precision, certainty, last update and update rate. Only [24] has investigated the role of imperfection in a more systematic way and identified the following types of imperfection: unknown values, contradictory values, imprecise values, and incorrect values. Feature values are further classified according to their source and persistence into sensed, static, profiled and derived. The causes of imperfection are analyzed along this classification.

Data management research on handling imperfection (for an overview see [25] or [26]) has explored probabilistic and other extension to traditional data models (e.g. [27] or [28]). However, none of them actually considers the time dimension und thus the aging problem in depth.

6   Conclusions and Outlook

6.1   Conclusions and Future Research

The proposed solution integrates the aspects of imperfection and dynamics into the context model and structures provides a layered structure similar to traditional data management applications. This allows for scalable solutions and appropriate decoupling of different management aspects. In addition to that it offers different interfaces to context-aware applications at different levels of complexity. Dedicated context sources or sophisticated context-aware services will access the context management infrastructure primarily on the internal layer, whereas applications that extend their traditional functionality with some context-aware features and provide the user only with limited interaction possibilities with the context information will primarily access the external layer. The logical layer is for added-value services that do not want to deal with the temporal perspective or inconsistent data.

Future research will explore the heuristic strategies used for inconsistency resolution and for subcontext switching via simulations. This will lead to insights which strategies are most suitable for specific domains characteristics.

6.2 Outlook

The concept of aging as presented here in order to realize a gradual forgetting of collected information is not limited to the area of user context data management, but can also be applied to a broader range of applications that have to rely on imperfectly sensed or inferred data whose validity is time-dependent. The awareness of the imperfection of the collected data can lead to more robust applications, which is especially interesting for knowledge-based systems like expert systems. The imperfection of machine learning results can be compensated that way.

If we generalize the aging mechanism to other information items (i.e., from the user context to the resource context), it would also be possible to enhance the relevance of results of information retrieval systems. Documents could be classified into different categories according to the maturity of their content (for the knowledge maturing process concept see [29]). It can be assumed that immature knowledge becomes outdated much faster than mature knowledge, which is expressed in text books, or e-learning courses. This way, the aging functions can be applied to decrease the probability of relevance of immature information objects.

With the layered approach presented in this paper, it is also possible to enhance Semantic Web application which are based on description logics formalisms with this aging functionality. Basically, aging is applied to the A-Box facts. This is especially beneficial if the A-Box is rather dynamic and learnt automatically based on heuristics. The implementation can be done computationally efficient in Datalog-based approaches for description logics reasoning like KAON2 [30]: the user context management can be plugged in as a provider for extensional predicates.

Acknowledgements

Part of this work was co-funded by the European Commission under contract IST-2001-32518 within IST Framework Programme 5.

References

1. Dey, A.K.: Understanding and using context. Personal and Ubiqutous Computing Journal **1** (2001) 4–7
2. Winograd, T.: Architectures for context. Human-Computer Interaction **16** (2001)

3. Schmidt, A., Winterhalter, C.: User context aware delivery of e-learning material: Approach and architecture. Journal of Universal Computer Science (JUCS) **10** (2004) 28–36

4. Schmidt, A.: Bridging the gap between knowledge management and e-learning with context-aware corporate learning solutions. In Althoff, K.D., Dengel, A., Bergmann, R., Nick, M., Roth-Berghofer, T., eds.: Post Conference Proceedings of the 3rd Conference on Professional Knowledge Management - Experiences and Visions (WM05), Springer (2005)

5. Elst, L., Abecker, A., Maus, H.: Exploiting user and process context for knowledge management systems. In: Workshop on User Modelling for Context-Aware Applications at UM 2001. (2001)

6. Schwarz, S.: A context model for personal knowledge management. In: Proceedings of the IJCAI-05 Workshop on Modeling and Retrieval of Context Edinburgh, July 31 - August 1, 2005. CEUR Workshop Proceedings (2005)

7. Haeberlen, A., Flannery, E., Ladd, A.M., Rudys, A., Wallach, D.S., Kavraki, L.E.: Practical robust localization over large-scale 802.11 wireless networks. In: Proceedings of the Tenth ACM International Conference on Mobile Computing and Networking (MOBICOM). (2002)

8. Strang, T., Linnhoff-Popien, C.: A context modeling survey. In: Workshop on Advanced Context Modelling, Reasoning and Management, UbiComp 2004 - The Sixth International Conference on Ubiquitous Computing, Nottingham/England. (2004)

9. Chen, H., Finin, T., Anupam, J.: Semantic web in the context broker architecture. In: PerCom 2004. (2004)

10. Schmidt, A.: Management of dynamic and imperfect user context information. In Meersman, R., Tari, Z., Corsaro, A., eds.: OTM Workshops. Volume 3292 of Lecture Notes in Computer Science., Springer (2004) 779–786

11. van Bunningen, A., Feng, L., Apers, P.M.: Context for ubiquitous data management. In: International Workshop on Ubiquitous Data Management (UDM2005). (2005)

12. Wang, X., Gu, T., Zhang, D., Pung, H.: Ontology based context modeling and reasoning using owl. In: IEEE International Conference on Pervasive Computing and Communication (PerCom'04), Orlando, Florida. (2004)

13. Strang, T., Linnhoff-Popien, C., Frank, K.: CoOL: A Context Ontology Language to enable Contextual Interoperability. In Stefani, J.B., Dameure, I., Hagimont, D., eds.: LNCS 2893: Proceedings of 4th IFIP WG 6.1 International Conference on Distributed Applications and Interoperable Systems (DAIS2003). Volume 2893 of Lecture Notes in Computer Science (LNCS)., Paris/France, Springer Verlag (2003) 236–247

14. Nebel, I., Smith, B., Paschke, R.: A user profiling component with the aid of user ontologies. In: Workshop Learning - Teaching - Knowledge - Adaptivity (LLWA 03), Karlsruhe. (2003)

15. Heckmann, D.: A specialized representation for ubiquitous computing and user modeling. In: First Workshop on User Modeling for Ubiquitous Computing, UM 2003. (2003)

16. Dolog, P., Nejdl, W.: Challenges and benefits of the semantic web for user modelling. In: AH2003 Workshop at WWW2003. (2003)

17. Lonsdale, P., Beale, R.: Towards a dynamic process model of context. In: Workshop on Advanced Context Modeling, Reasoning and Management, Ubicomp 2004. (2004)

18. Kazakos, W., Nagypal, G., Schmidt, A., Tomczyk, P.: Xi3 - towards an integration web. In: 12th Workshop on Information Technology and Systems (WITS '02), Barcelona, Spain (2002)

19. Jalkanen, J.: User-initiated context switching using nfc. In: Proceedings of the IJCAI-05 Workshop on Modeling and Retrieval of Context Edinburgh, July 31 - August 1, 2005. CEUR Workshop Proceedings (2005)

20. Schmidt, A.: Context-steered learning: The learning in process approach. In: IEEE International Conference on Advanced Learning Technologies (ICALT '04), Joensuu, Finland, IEEE Computer Society (2004) 684–686

21. Maedche, A., Motik, B., Stojanovic, L., Studer, R., Volz., R.: An infrastructure for searching, reusing and evolving distributed ontologies. In: Proceedings of WWW 2003, Budapest, Hungary. (2003)

22. Buchholz, T., Küpper, A., Schiffers, M.: Quality of context: What it is and why we need it. In: 10th International Workshop of the HP OpenView University Association (HPOVUA 2003), Geneva, Switzerland. (2003)

23. Judd, G., Steenkiste, P.: Providing contextual information to ubiquitous computing applications. In: 1st IEEE Conference on Pervasive Computing and Communication (PerCom 03), Fort Worth. (2003) 133–142

24. Henricksen, K., Indulska, J.: A software engineering framework for context-aware pervasive computing. In: PerCom, IEEE Computer Society (2004) 77–86

25. Motro, A.: Sources of uncertainty, imprecision and inconsistency in information systems. In Motro, A., Smets, P., eds.: Uncertainty Management in Information Systems: From Needs to Solutions. Kluwer Academic Publishers (1996) 9–34

26. Parsons, S.: Current approaches to handling imperfect information in data and knowledge bases. IEEE Transactions on Knowledge and Data Engineering **8** (1996) 353–372

27. Barbará, D., García-Molina, H., Porter, D.: The Management of Probabilistic Data. ACM Transactions on Knowledge and Data Engineering **4** (1992) 487–502

28. Fuhr, N., Rlleke, T.: A probabilistic relational algebra for the integration of information retrieval and database systems. ACM Transactions on Information Systems **15** (1997) 32–66

29. Schmidt, A.: Knowledge maturing and the continuity of context as a unifying concept for knowledge management and e-learning. In: Proceedings of I-KNOW 05, Graz, Austria. (2005)

30. Hustadt, U., Motik, B., Sattler, U.: Reducing shiq-description logic to disjunctive datalog programs. In: Principles of Knowledge Representation and Reasoning: Proceedings of the Ninth International Conference (KR2004), Whistler, Canada, June 2-5, 2004. (2004) 152–162

Designing the Context Matching Engine for Evaluating and Selecting Context Information Sources

Maria Chantzara and Miltiades Anagnostou

Computer Networks Laboratory, School of Electrical & Computer Engineering,
National Technical University of Athens (NTUA),
9 Heroon Polytechniou Str, Zografou 157 73, Athens, Greece
{marhantz, miltos}@telecom.ntua.gr

Abstract. The easy creation of context-aware services requires the support of management facilities that provide ways to more easily acquire, represent and distribute context information. This paper claims that the quality level of a context-aware service determines the context information to be obtained. On the other hand, using context data produced by unsteady sources may affect the users' satisfaction. In this perspective, we introduce the Context Matching Engine that trades off the cost, the user preferences and the quality of the available context information in order to discover the best context sources for each customized context-aware service. According to the proposed approach, there is no need for the services to know beforehand the context providers to retrieve information, but the evaluation and the quality-aware selection of the context information on context request are envisioned. Finally, it allows services to be ported easily to environments with different set of context sources.

1 Introduction

The advances in wireless communications and user mobility have given quite a boost to the research about new classes of applications, namely the Context-Aware Services (CASs) that get aware of the execution environment such as location, time, user's activities, devices' capabilities in order to tune their intended functionalities and adapt to the changing environment and user requirements [1]. The development and provisioning of CASs need to be supported by management facilities capable of collecting, manipulating, reasoning and disseminating context information. In this respect, researchers have been building tools and architectures to tackle these issues and facilitate the easy creation of CASs.

As research on context-awareness evolves, new context sources are expected to come and go rapidly, and context requests will increase. To make matters worse, among the available context information sources, there are sets of sources that provide information referring to the same entity, but are produced based on different sensing technology or/and processing technique under the administration of different business entities. As a result, even though the information may refer to the same entity, it varies in its update frequency, its accuracy, its format of representation, and its price. Moreover, the sensed context information can be inaccurate or unavailable due to

T.R. Roth-Berghofer, S. Schulz, and D.B. Leake (Eds.): MRC 2005, LNAI 3946, pp. 101–117, 2006.

sensor or connectivity failures, while user-supplied information is subject to human error and staleness. The reliance of CASs to imperfect information can often cause usability problems [2] threatening service viability. As a mean to judge the reliability of context, the notion of Quality of Context has been proposed and the reasons it is needed are analyzed [3]. The context models are designed to tag quality attributes to the context data [4] and the CASs define quality attributes as properties of the information to be collected [5].

On the other hand, CASs are characterized by a higher level of mobility since the users tend to move among different heterogeneous environments where different context providers operate and provide information. Even though many research efforts [6] have been conducted to specify and materialize frameworks and toolkits for context-awareness, these efforts lack to deal with the provisioning of the same CASs in multiple environments that reveals the necessity to elaborate on issues that have to do with the collection of data from multiple context sources maintained by different administrative entities [7].

Furthermore, the provisioning of the same context information with different characteristics calls for advanced personalization features that can lead to the provisioning of sophisticated CASs. Specifically, during subscription the user enters the parameters that personalize the service operation. These parameters form the user's service profile that encompasses his/her needs from the CAS. For example, a user that subscribes to a context-aware restaurant finder service could specify that he/she wishes the weather conditions to be taken into account in contrast to another user that just wants to find what are the restaurants in the city that match his/her taste preferences. In addition to the profile, the user identifies the price ceiling for the personalized service provisioning and awaits best possible service delivery within his/her price range [8]. From service management point of view this connotes a complete inversion: not only the service defines the price but also the price causes the type of service provisioning. Finally, both the price ceiling and the subscription parameters determine the quality level of offered service and the information to be obtained.

As a consequence of the aforementioned issues, a new challenge referring to the alignment of the context information that is used by a CAS with the customized service objectives is introduced. In the perspective of this challenge, we have elaborated on a novel approach for the evaluation and selection of context providers that balances quality of information, economic cost and user preferences. This approach requires the enhancement of context management frameworks with the *Context Matching Engine* (CME) that performs the proposed multicriteria decision making. Thereafter, there is no need to know beforehand the providers to obtain the required information, but the quality-aware and cost-driven discovery of the context providers is envisioned. The proposed approach also enables flexibility to failures of context providers or appearance of new ones.

The paper is organized as follows: In Section 2 the context and CAS provision model are described. In Section 3, the Context Matching Engine is briefly presented, while in section 4 the functionality of the Context Matching Engine is detailed. The evaluation of the proposed approach in terms of a specific CAS and the comparison to alternative approaches is reported in section 5. Finally, Section 6 provides some conclusive remarks and future plans.

2 Context and Context-Aware Service Provision Model

Context information goes through four lifecycle phases till a CAS can access it. The first phase is context sensing that represents the acquisition of context from the Context Information Sources e.g. sensors, devices, data bases, network. Until the CASs can actually use the data, context processing and dissemination take place. Context processing includes the generation of high-level context information from primitive data by enforcing functions of interpretation, filtering, aggregation and inference. Context dissemination refers to the efficient distribution of context to the CASs. The context lifecycle ends with the context usage that describes the utilization of context by CASs in order to trigger the appropriate actions. The aforementioned phases introduce new opportunities in the market arising out of the exploitation of context information trade. The different business roles participating in CAS provision are depicted in Fig.1. A CAS Provider is responsible for providing CASs to the customers/users and managing them. The Context Provider deploys and operates the various Context Information Services that communicate with the sources and sense or process information. Even though a Context Provider could operate more than one context source, due to simplicity reasons, we assume that each Context Provider operates only one source for the rest of the paper. Finally, the Context Broker is responsible for handling the distribution of context data to the CAS Providers or directly to the CASs. This role provides efficiency and reduces time-to-market for CASs. CASs could interact with the Context Providers to acquire the necessary information, but this would require more programming effort for the CAS developers.

The CONTEXT project [9] has successfully used this model by designing and implementing the Context Broker to coordinate and facilitate communication between context providers and CASs. According to the CONTEXT design, the Context Broker offers the *Context Information Provider Interface* that enables context providers to declare and supply the context information they produce and the *Context Information User Interface* that enables CASs to request the information they want to consume.

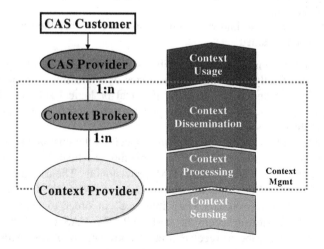

Fig. 1. Context phases and involved actors

The distributed Context Brokers act as a federation that cooperates to answer context requests. The CASs and the context sources are using a common model to describe context information, which can be viewed as a kind of primordial ontology. In specific, we have introduced the notion of Context Modules, which are Extensible Markup Language (XML) descriptions consisting of three parts: the context name, the input and the output parameters. Each input and output parameter consists of three parts: the name, the type (e.g. integer, string, list) and the parameter value. A context source, registering context information, provides the context module description in terms of context name and the names of input and output parameters. It also provides the values of the identification parameters which are a subset of the Context Module input parameters; the values of the rest of them are specified by the context consumer on context request. A context consumer, requesting specific context information, provides the name of the context module, the input parameters (names and values) and the output parameter names it wishes to acquire. The described data-centric approach does not require the CASs to name a specific context provider from which they wish to retrieve some information but just the information they wish to obtain. Finally, the Context Broker is responsible for discovering and then interacting with the appropriate information producer and retrieving the values corresponding to the requested output parameters.

The context providers utilize the "Context Information Provider Interface" offered by the Context Broker to register the type of information they provide accompanied with the data quality attributes (DQ) that characterize it. The different nature of the context information complicates the identification of general quality attributes, while many different lists of attributes are presented in literature [10]. Considering work in [3],[11], we conclude to the following list that characterize context information and may affect the context-aware service behavior:

1. *Accuracy*: It describes how the information depicts the reality, and depends on the sensing or processing technique that is used as well as the internal sensing problems that may occur.
2. *Refresh rate*: It describes the time period that a new measurement of the specific data attribute is done.
3. *Time sample*: It is also known as timestamp and describes the time point that the value has been obtained.

It is obvious that the complete list of attributes does not refer to all types of context information. Specifically, it mostly characterizes dynamic information that changes frequently over time and is produced by physical or logical sensors. On the other hand, the quality of static information such as user's profiles and preferences is determined by the time sample.

At CAS subscription phase, the CAS user specifies the customization parameters that form the user's service profile (SP) and encompasses his/her requirements. These parameters point to the quality level of CAS provisioning. The user also specifies the price ceiling for the personalized service provisioning. When the CAS is executed, it utilizes the "Context Information User Interface" in order to declare the context information to be retrieved according to the SP. The CAS could either query for context values or subscribe to receive context notifications. Along with the context information to be collected, the service indicates the constraints for the context provisioning that will be described in a following section. In response to the context request, the Context Broker discovers the context sources to acquire the information, contacts them and the required data are retrieved. The context sources deliver the requested piece of information accompanied with its quality characteristics in order that the information is verified. Furthermore, the service is informed about the information characteristics in order to adapt its behavior accordingly e.g. inform the user that the context information is possibly inaccurate or imprecise. The timestamp describing the production time of the data is also provided at this time point.

3 The Context Matching Engine

As it is analyzed in the introduction, the richness of context information gathered from sensors and human users as well as the need to execute the CASs in multiple environments require decision-making regarding the exact information that a CAS should use. In order to tackle these issues, we claim that the descriptive naming of context sources [12] is not sufficient but a more powerful design is needed. In this perspective, we introduce the Context Matching Engine (CME), operating in each domain under the administration of the Context Broker. The CME is responsible for discovering per customized CAS the appropriate context sources that fulfill the service's and the user's objectives. The logical decomposition of the CME is depicted in Fig. 2.

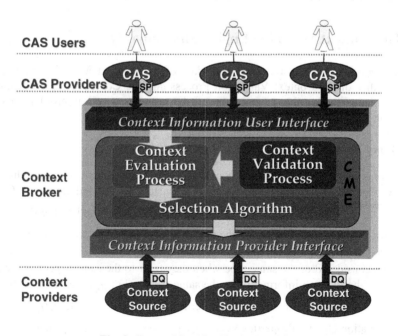

Fig. 2. Context Matching Engine Modules

As the figure shows, the CME accommodates the following modules:

1. The *Context Validation Process* that monitors the behaviour of the context providers and estimates their fidelity and the time response for answering context requests. It also detects failures or changes of the context sources.
2. The *Context Evaluation Process* that considers the selected service profile, the user subscription data, and the outcomes of the context validation process and computes the expected service satisfaction for the available context information.
3. The *Context Selection Algorithm* that determines the appropriate information to be utilized by the CASs given as input the results of the previous processes.

When a CAS contacts the Context Broker asking for specific information, the CME undertakes the task of deciding upon the appropriate context sources to retrieve the data. The Context Evaluation Process is initially triggered to evaluate the available context sources on behalf of the user. The Selection Algorithm is then executed to determine the optimal combination of sources. Based on this decision the sources are contacted and the data are retrieved. During the CAS operation, the Context Validation Process monitors the quality of the provided information. This facilitates the optimal context source selection for serving the next context requests as well as the current ones. In this respect, autonomic features are exhibited through the switch of the context sources when new sources are registered or one of the already selected fails or one of the non selected is upgraded. In the event of a context source change, the Context Evaluation Process and the Selection Algorithm are again triggered to decide on the rebinding to another source.

4   The Context Matching Engine Functionality

In this section, after defining the input data of the Context Matching Engine, the functionality of the Context Validation Process and the Context Evaluation Process are detailed. Finally, the formulation of the context selection problem and the Selection Algorithm that solves it are presented.

4.1   Input Data

The market of context information consists of the set of context providers that offer context data. Every context provider CP_{ij} registers to the federation of Context Brokers the produced information I_{ij} that is of type $i \in (1, N)$ and of the data quality $j \in (1, M_i)$. N is the total number of the available context items, while M_i denotes the total number of the available quality levels of context items of type i. Two context items are considered of the same type when they refer to the same aspect of the same entity, e.g. the location information of the same person. For I_{ij}, the quality attributes, depicted in Table 1, are defined. Therefore, for each I_{ij}, $i \in (1, N)$, $j \in (1, M_i)$ we define the Information Quality Properties Vector $IQPV_{ij} =< A_{ij}, Tr_{ij}, Ts_{ij} >$. P_{ij} is the purchase price for I_{ij}.

Table 1. Quality Properties of Context Information

Quality Property Symbol	Quality Property Description
A_{ij}	Accuracy of I_{ij} measured in the same metric units as I_{ij}
Tr_{ij}	Refresh rate of I_{ij} measured in time metrics
Ts_{ij}	Time sample of I_{ij} measured in date and time metrics

Table 2. Input Data

		Input Data	Input Attribute Symbol
Context Information Market		Information Quality Properties Vector of information I_{ij}	$IQPV_{ij} =< A_{ij}, Tr_{ij}, Ts_{ij} >$
		Cost of purchasing information I_{ij}	P_{ij}
		L specific types of context items to be obtained	$\{k_1, k_2, ..., k_L\} \subseteq (1, N)$
Service Profile/ User Subscription Data		For each required context item the Acceptable Properties Vector	$APV_{k_z} = \left\langle A\max_{k_z}, Tr\max_{k_z}\right\rangle,$ $\forall k_z, z \in (1, L)$
		For each required context item the Properties Weight Vector	$PWV_{k_z} = \left\langle a_{k_z}, b_{k_z}, c_{k_z}, d_{k_z}, e_{k_z}\right\rangle,$ $\forall k_z, z \in (1, L)$
		Maximum cost of purchasing required information	$P_{\max}$
		Latency bound for obtaining the information	$T_{\max}$

When the CAS customer subscribes to the CAS, he/she selects one of the predefined service profiles and specifies the customization parameters. The selected profile requires the acquisition of L specific context items of $\{k_1, k_2, ..., k_L\} \subseteq (1, N)$ type. For each $k_z, z \in (1, L)$ requested context item the Acceptable Properties Vector $APV_{k_z} = \left\langle A\max_{k_z}, Tr\max_{k_z}\right\rangle$ is specified. Moreover, the Properties Weight Vector $PWV_{k_z} = \left\langle a_{k_z}, b_{k_z}, c_{k_z}, d_{k_z}, e_{k_z}\right\rangle$ weights the importance of accuracy, timeliness, refresh rate, and the source's time response and fidelity respectively, conforming to: $a_{k_z} + b_{k_z} + c_{k_z} + d_{k_z} + e_{k_z} = 1$. Both APV and PWV are determined by statistics produced by the previous service provisioning. Moreover, the CAS customer specifies the amount of money that the specific CAS usage should cost him/her. Based on this value, the maximum cost $P_{\max}$ of purchasing the necessary context information is calculated. Furthermore, the selected SP suggests the latency bound $T_{\max}$ describing the bound of the time units for obtaining the information. Additionally, the CAS specifies the policies describing how the CME should behave in case the requirements can not be satisfied. The input data of the problem are summarized in Table 2.

4.2 The Context Validation Process

Since it is not fair to expect that the context providers advertise reasonably well the corresponding data quality properties, the Context Validation Process monitors the status of context information provisioning between the context sources and the services and provides an objective measurement of the trustworthiness of sources based on a learning model. It finds out failures or just changes of the context sources as well as the registration of new ones that could be in the interest of the customized CASs that operate currently, and triggers their evaluation so that possible switch of the context sources is performed. It also reviews whether the quality attributes accompanying the delivered context values are aligned with the registration parameters that the context provider has specified. Thus, it examines the freshness of the delivered information by checking whether the actual refresh rate harmonizes with the refresh rate that the source has registered. Moreover, it measures the response time of the source to the context requests in order to determine the freshness of the delivered information. Finally, the cross-validation of the context values provided by different context sources would be an interesting enhancement.

Therefore, the Context Validation Process determines some quality attributes that characterize the providers of the information and change with time, current network topology and characteristics. These are:

1. *Time response:* It corresponds to the time interval that is required by the information producer to deliver the context data. It is particularly useful for time criticality issues.
2. *Fidelity:* It is the probability that the quality attributes advertised at registration match the actual ones. Accordingly, those producers that have not been detected as incorrect tend to be more trusted by the consumers and the fidelity is higher. This parameter allows the Context Broker to choose on behalf of the service how much risk it willing to take in the hope of receiving good quality information. The fidelity is calculated as follows: Considering the time point T after the context provider CP_{ij} delivers the latest context value. The Current Fidelity $CF_{ij}(CP_{ij},T)$ takes the value "1" if the latest provided value is according to the registration parameters; otherwise it takes the value "0". We also define the parameter W called window describing the number of the past refresh cycles that we monitor. As W gets higher, the estimation of fidelity considers more of the past behavior of the source but requires more resources. Finally, the Fidelity at time point T is the average of the CF_{ij} that occur during the previous w refresh cycles:

$$F(CP_{ij},T) = \frac{1}{W} * \sum_{t=T-W*Tr_{ij}}^{t=T} CS(CP_{ij},t) \tag{1}$$

To sum up, for every context provider CP_{ij} the Context Validation Process calculates the attributes that are given in Table 3. These form the Context Provider Quality Properties Vector $CPPV_{ij} = < Tresp_{ij}, F_{ij} >$.

Table 3. Quality Properties of Context Information Providers

Quality Property Symbol	Quality Property Description
$Tresp_{ij}$	Time response of CP_{ij} measured in time metrics
F_{ij}	Fidelity of CP_{ij} is a probability

4.3 The Context Evaluation Process

The evaluation of the available context items for a specific service and service profile is determined based on a utility model that quantifies the expected user and service's satisfaction when they are used. According to it, we define a function that takes input the user and service requirements and computes the usability of the context items in terms of actuality and freshness. This function is called Utility Function $U(I_{ij})$ and its output is called Utility. It is comprised by the five factors: $U_A, U_T, U_{Tr}, U_{Tresp}, U_F$, that measure the utility in terms of accuracy, timeliness, refresh time, and producer's time response and fidelity, respectively. The U_A, U_F express actuality, while the U_T, U_{Tr}, U_{Tresp} express freshness. The utility increases as the context item is more accurate, as it is quicker available to the context consumer, and as it remains valid for a longer period. Regarding accuracy and refresh rate, the utility of a context item increases as the properties A_{ij}, Tr_{ij} decrease respectively. The formulation of the factors for information I_{ij} considers the quality constraints imposed by the selected service profile, namely the $A\max_i, Tr\max_i$, and the time point $Tcurrent$ when the context request is issued. Ts_{ij} is the timestamp of the latest produced context value for I_{ij}. The factors take values in $[0,1]$ and their formulas are listed below:

- Measuring information's utility in terms of accuracy:

$$U_A(I_{ij}) = 1 - \frac{A_{ij}}{A\max_i} \qquad (2)$$

- Measuring information's utility of dynamic and static context information in terms of timeliness respectively:

$$U_T(I_{ij}) = 1 - \frac{Tcurrent - Ts_{ij} + Tresp_{ij}}{Tr_{ij}} \qquad (3)$$

$$U_T(I_{ij}) = 1 - \frac{Tcurrent - Ts_{ij}}{\max_{k=i, j=1}^{k=i, j=M_i} (Tcurrent - Ts_{kj})} \qquad (4)$$

– Measuring information's utility in terms of producer's time response:

$$U_{Tresp}\left(I_{ij}\right) = 1 - \frac{Tresp_{ij}}{T_{max}} \tag{5}$$

– Measuring information's utility in terms of refresh rate:

$$U_{Tr}\left(I_{ij}\right) = 1 - \frac{Tr_{ij}}{Tr\,max_i} \tag{6}$$

– Measuring information's utility in terms of producer's fidelity:

$$U_F\left(I_{ij}\right) = F_{ij} \tag{7}$$

Finally, the Utility Function of I_{ij} taking into account the ranking of the utility factors as suggested by the selected service profile $PWV_i = \langle a_i, b_i, c_i, d_i, e_i \rangle$ is formulated as follows:

$$U\left(I_{ij}\right) = a_i U_A\left(I_{ij}\right) + b_i U_T\left(I_{ij}\right) + c_i U_{Tr}\left(I_{ij}\right) + d_i U_{Tresp}\left(I_{ij}\right) + e_i U_F\left(I_{ij}\right) \tag{8}$$

In case a quality attribute is not available for the specific context item, the corresponding utility factor equals zero. For example, for the static information the utility function depends only on the timeliness of the information, the response time and the fidelity of the producer.

4.4 Selection Problem Formulation and Selection Algorithm

Having described the parameters of the problem, a formal statement of the selection problem is as follows: "Given a market of context information consisting of the set of context providers CP_{ij}, $(i \in (1, N), j \in (1, M_i))$ that each sells context information I_{ij}, the corresponding characteristics: $IQPV_{ij}$ and $CPPV_{ij}$, the cost P_{ij} for obtaining information I_{ij}, a finite number of L context items of distinct types: $\{k_1, k_2, ..., k_L\} \subseteq (1, N)$ that need to be obtained for the customized CAS, the constraints imposed by the selected service profile: APV_{k_z}, PWV_{k_z} $\forall k_z, z \in (1, L)$ as well as the maximum acceptable amount of money P_{max} that can be spent and the maximum latency T_{max} for obtaining the required context information, the objective is to determine the context sources that will provide the required information so as to maximize the total utility under the imposed constraints."

Taking into account the formal description of the problem, the objective function that has to be maximized is the following:

$$UF = \sum_{i=k_1}^{i=k_L} \sum_{j=0}^{j=m} U\left(I_{ij}\right) x_{ij} \tag{9}$$

$x_{ij} \in \{0,1\}$ is the decision variable describing if the information of type i is being acquired from the context producer CP_{ij} or not. Moreover, the following constraints should be met in order to find the optimal solution:

$$\sum_{i=k_1}^{i=k_L} \sum_{j=0}^{j=M_i} P_{ij} x_{ij} \leq P_{\max} \tag{10}$$

$$\sum_{i=k_1}^{i=k_L} \sum_{j=0}^{j=m} Tresp_{ij} x_{ij} \leq T_{\max} \tag{11}$$

$$\max_{\substack{i=k_1}}^{i=k_L} \left\{ \sum_{j=0}^{j=M_i} Tresp_{ij} x_{ij} \right\} \leq T_{\max} \tag{12}$$

$$\forall i \in \{k_1, k_2, ..., k_L\}, \sum_{j=0}^{j=M_i} x_{ij} = 1 \tag{13}$$

$$\forall i \in \{k_1, k_2, ..., k_L\}, \sum_{j=0}^{j=M_i} A_{ij} x_{ij} \leq A \max_i \tag{14}$$

$$\forall i \in \{k_1, k_2, ..., k_L\}, \sum_{j=0}^{j=M_i} Tr_{ij} x_{ij} \leq Tr \max_i \tag{15}$$

Equations (10), and either (11) or (12) express the fulfillment of the CAS's requirements regarding the cost and latency for the information acquisition respectively. The equation (11) holds for the cases that context items are dependent and should be obtained sequentially, while the equation (12) holds for the cases that the context items are independent and can be obtained in parallel. Moreover, the equation (13) ensures that for each required context item, only one information source is used. The equations (14) and (15) are filtering the available items in terms of the APV. Finally, the formulation of the problem is aligned to user's behaviour when he/she decides the goods to purchase. Thus, users possesses a fixed amount of money for purchasing an item of specific type and pick the item with the highest quality, while providers set a minimum quality level for the item to buy and try to maximize the profits. Regarding the case that no feasible solution can be found due to constraints (10) and/or (12) or (13), the service policies are considered in order to relax the relevant constraint, namely increase $P_{\max}$ or $T_{\max}$ respectively. In case that there is no available context item satisfying (14) or (15), the CME selects the one that is expected to provide the best possible user satisfaction, namely the one that minimizes the violation of (14) and (15).

An important class of combinatorial optimization problems is the *Knapsack Problem* and its variants [13]. Numerous problems of different fields such as capital budgeting, cargo loading and resource allocation are modeled as a variant of the knapsack problem. The objective of the original Knapsack Problem is to optimize resource allocation, or more precisely, how to distribute a fixed amount of resources among several actions in order to obtain a maximum payoff. The problem of context provider selection that has been formulated belongs to the category of *Multi-Choice Multi-Dimensional Knapsack Problem* (MMKP) [14],[15]. The definition of MMKP is: There are n groups of items. Group i contains l_i items. Each item has a particular value and consumes specific amount of the m resources. The objective is to pick exactly one item from each group in order to achieve maximum total value of the collected items, subject to the m resource limits. In the case of the context provider selection problem, each group represents a requested type of information while the items in it represent the available context items. The resource constraints refer to the cost and the latency for obtaining the information. Finally, the value that needs to be maximized is the total utility of the picked context items. QoS management problem in multimedia systems is a well known problem that is also formulated as MMKP [16]. But this is an easier instance than our problem because the values follow monotone feasibility order in relation to resource consumption.

For the current implementation of the CME, we are using the *Maximizing Value per Resource Consumption* (MVRC) heuristic (For instances of n groups of items, l items per every group and m resource constraints, the computational complexity is $O(nml+nl!+m+n^2(l-1)m+n^2(l-1)^2m))$. The MVRC is an improvement of the algorithm HEU [17] but solves the problem in less time that is important for cases that require real-time decision making like ours. Unlike the HEU that tries to minimize the resource consumption, the MVRC picks the items with the maximum value per aggregated resource consumption. Comparing the MVRC with the exhaustive approach showed that the MVRC produces solutions with 100% optimality for small data sets (small number of groups, items per group and resource constraints) and 85-97% optimality for large data sets. Being beyond the scope of this paper, the further details and evaluation results of the heuristic are not presented.

5 Evaluation

5.1 Demonstration Service

In our future research plans, we intend to implement a context aware restaurant finder service like the one described in [3], which may consider user preferences, user location, weather conditions, traffic conditions, restaurant menus in order to advice the user about the best available restaurant to have lunch. However, in order to demonstrate the proposed approach, we have performed some simulations of a simpler context-aware restaurant finder service. This CAS considers only the user's preferences along with user's location in order to determine the appropriate restaurant for the user to have lunch. In case that the user would also like weather conditions, namely temperature information, to be also taken into account, the service retrieves location information of the user and based on this data, it also collects the temperature information. In Table 4, the properties of the offered location and temperature information are shown. We assume that all sources have started producing values when the time was T=0sec and the new values are produced according to the registered refresh rate. Moreover, the CAS requests information when time is $T_{current}=7sec$.

The CAS provider specifies the following weights regarding the gathering of location and temperature information for the specific CAS: $PWV_1=\{0.5,0.1,0.1,0.2,0.1\}$ and $PWV_2=\{0.1,0.1,0.3,0.5,0.1\}$. Regarding the location information, accuracy is of higher

Table 4. Context Information Market

	Location Information (m)						Temperature Information (°C)					
	I_{11}	I_{12}	I_{13}	I_{14}	I_{15}	I_{16}	I_{21}	I_{22}	I_{23}	I_{24}	I_{25}	I_{26}
P	150	100	50	80	190	20	50	70	80	100	130	40
A	5	100	900	200	1	1000	2	1,5	1	1	0,5	3
Tr (sec)	10	60	240	100	6	220	10	8	10	5	5	10
Ts (sec)	5	0	0	0	7	0	6	6	7	7	7	6
Tresp(sec)	0,9	2,1	1	0,5	1,2	1,2	0,5	1	1,9	2,5	1	3,5
F (%)	100	100	100	100	100	100	100	100	100	100	100	100

Table 5. Utility Calculation for User A

	Location Information						Temperature Information					
	I_{11}	I_{12}	I_{13}	I_{14}	I_{15}	I_{16}	I_{21}	I_{22}	I_{23}	I_{24}	I_{25}	I_{26}
U_A	0.995	0,9	0,1	0,8	0,999	0	0,333	0,5	0,66	0,66	0,833	0
U_T	0.71	0,848	0,966	0,925	0,8	0,963	0,85	0,75	0,81	0,5	0,8	0,55
U_{Tr}	0,958	0,75	0	0,583	0,975	0,083	0	0,199	0	0,5	0,5	0
U_{Tresp}	0.775	0,475	0,75	0,875	0,7	0,7	0,875	0,75	0,525	0,375	0,75	0,125
U_F	1	1	1	1	1	1	1	1	1	1	1	1
U	0,919	0,804	0,396	0,825	0,917	0,344	0,659	0,660	0,510	0,554	0,788	0,217

importance while regarding temperature information time specific properties are more important. We consider two users A and B: User A wishes that temperature information is taken into account while he/she does not have any special requirements regarding the restaurant's location. In this case, the CAS is expected to work well if it retrieves user's location information with low quality as well as the temperature information of the town that hosts the user. On the other hand, User B wishes to find the nearest restaurant that matches his/her preferences. Therefore, for this user the service requires user's location information with greater detail. The two users select the corresponding profiles that match their needs. The imposed constraints are: $APV_1=$ {1000, 240}, $APV_2= \{3,10\}$ for User A and $APV_1= \{5,20\}$ for User B. The price ceilings that are defined by each user are $P_{max}= 100$, 170 respectively, while the maximum latency is $T_{max}=4sec$. Table 2 shows the expected utility per context item for User A.

For User A, our approach concludes to the acquisition of information I_{13} and I_{21}. In this case, the objective function (9) is maximized with value $0,396+0,659=1,056$ while all constraints are also satisfied. This solution costs 100. If we only considered the constraints APV_1, APV_2 to perform the selection, the information that would be used is I_{16} and I_{26}, achieving total quality $0,344+0,217=0,562$ with cost 60. However, not only has this solution lower total utility, it does not satisfy constraint (11) as well.

For User B, only I_{11} and I_{15} satisfy (14) or (15) constraints, and represent feasible solutions. Our approach decides that the optimal choice is information I_{11} with utility 0,336 and cost 150. The highest utility is 0,79 and is achieved by I_{15}. However, this is not a feasible solution because it costs 190.

5.2 Comparative Evaluation

In this section, we present the evaluation of the solution found by the proposed approach in relation to the solutions found by applying some alternative selection mechanisms. The analysis is based on the customized context-aware restaurant finder service provided to User A that was presented in the previous subsection. The selection mechanisms that we consider are:

1. *Context Matching Engine* (CME) that this paper proposes.
2. *Match Data Properties* (MDP) that selects the information conforming to the requirements related to APV, without considering resource constraints.
3. *Match Data Properties & Constraints* (MDP_C) that selects the information conforming to the requirements related to APV and the resource constraints.
4. *Match Cost* (MC) that selects the information harmonizing to the cost constraint, namely the context items that all together have the closest cost to P_{max}.
5. *Match Latency Bound* (MLB) that selects the information harmonizing to the latency constraint, namely the context items that can be delivered within the closest time to T_{max}.
6. *Maximum Total Utility* (MEU) that selects the information maximizing the total utility without considering any resource constraint.
7. *Context Matching Engine & No Weights* (ME_NW) that selects the information that matches the criteria of the APV and the two resource constraints and maximizes the expected utility. However, the ranking of the utility factors is not considered.

The Table 6 depicts the different selection mechanisms in relation to the involved user/services' requirements. The symbol "✓" means that this requirement is taken into account, while the opposite is depicted with the symbol "-".

Table 6. Selection Mechanisms

SELECTION MECHANISMS	Cost Constraint	Latency Constraint	Acceptable Data Properties	Utility Factors Ranking
CME	✓	✓	✓	✓
MDP	-	-	✓	-
MDP_C	✓	✓	✓	-
MC	✓	-	✓	-
MLB	-	✓	✓	-
MEU	-	-	✓	✓
CME_NW	✓	✓	✓	-

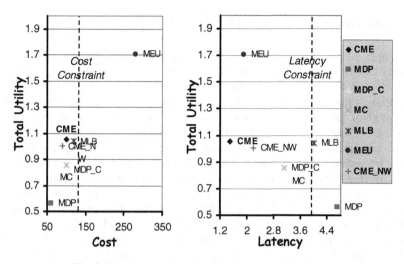

Fig. 3. Solutions obtained by the selection mechanisms

We applied the different selection mechanisms to the specific problem instance and recorded the obtained solutions in the diagrams of Fig. 3. These diagrams show thetotal utility of the solution in relation to the corresponding required cost and latency bound for retrieving the data.

Even though the previous diagrams are produced for the specific problem instance, we come to some conclusions that can be generalized. In specific, the MDP, which is used by the state-of-the-art context management frameworks and requires the smallest computation time since it decides about each required context item independently (for n group of context items each containing l items, MDP has O(nl)), generates solutions with the lowest total utility that possibly violate the resource constraints. Nevertheless, when resource constraints are considered in MDP_C, the generated solutions do not violate constraints, but they still have low total utility. On the other hand, the solutions of CME and CME_NW do not violate the resource constraints, and they have higher utility as well. Regarding the resource constraints, it can be observed that it not sufficient to consider only one resource. Considering only one constraint may produce solutions that violate the other. This is related to the fact that there is usually no correlation between the cost and the latency as well as correlation between the utility and the resource requirements of items. Regarding the ranking of the utility factors, we observe that the solutions produced by the mechanisms that consider them have significantly better utility.

In conclusion, by applying the proposed approach and the other different selection mechanisms to the service scenario we performed the initial evaluation of the CME functionality. This showed that, even though it rises up time and computation requirements for context provisioning, the enhancement of context management frameworks with the CME would improve both service's performance and user's experience from service usage. These are very important for the offering of viable and successful CASs.

6 Conclusions and Future Plans

This paper proposed a novel approach for quality-aware discovery of the context information. This approach supported by the Context Matching Engine, facilitates the seamless and transparent provision of dynamically changing context information regardless of the diverse user requirements and the changing system conditions. The description of the proposed model and an initial evaluation of it have been presented. New context sources can be added on-the-fly, while the chosen approach also caters for the provision of different versions of the same information type, with different update frequency, accuracy, and price, an especially useful feature for highly mobile users, moving between different heterogeneous environments as well as unsteady context sources. Autonomic features are exhibited through monitoring the context provision and rebinding to the best available context sources over the lifetime of the context requests. Our plans for future work in this area include the enhancement of the selection algorithm to be applicable for more complex context requests that do not conform to the sequential or in parallel context retrieval model as well as in cases that the context to be retrieved is determined during service operation, the development of an interactive user-specific service profile composition, and an intelligent context quality monitoring mechanism for critical applications. Finally, we intend to evaluate the solutions produced by our model and demonstrate its benefits utilizing a real-life context-aware service with real users.

References

1. Dey, A.: Understanding and Using Context. Personal and Ubiquitous Computing Journal. (February 2001). Vol. 5. No. 1. pp.4-7
2. Henricksen, K., Indulska, J.: Modelling and Using Imperfect Context Information. In Proceedings of the Workshop of the Advanced Context Modelling, Reasoning and Management associated with the Sixth International Conference on Ubiquitous Computing (UbiComp 2004). Nottingham/England. (September 2004)
3. Buchholz, T., Küpper, A., Schiffers, M.: Quality of Context: What it is and Why We Need It. In Proceedings of the Workshop of the HP OPenView University Association 2003 (HPOVUA 2003). Genevy. (July 2003)
4. Henricksen, K., Indulska, J., Rakotonirainy, A.: Modeling Context Information in Pervasive Computing Systems. In Proceedings of the 1st International Conference on Pervasive Computing, Pervasive 2002. Springer-Verlag. Vol. 2414, pp. 169-180
5. Judd, G., Steenkiste, P.: Providing Contextual Information to Pervasive Computing Applications. In Proceedings of the 1st IEEE International Conference on Pervasive Computing and Communications (PerCom'03)
6. Xynogalas, S., Chantzara, M., Sygkouna, I., Vrontis, S., Roussaki, S., Anagnostou, M.: Context Management for the Provision of Adaptive Services to Roaming Users. IEEE Wireless Communications. (April 2004). Vol. 11. No. 2. pp. 40-47
7. Meneses, F.: Context Management for Heterogeneous Administrative Domains. A. Ferscha et al. (eds.) Advances in Pervasive Computing. Austrian Computer Society. Austria. (2004)

8. Hegering, H., Kupper, A., Linnhoff-Popien, C., Reiser, H.: Management Challenges of Context-Aware Services in Ubiquitous Environments. In: M.Brunner, A. Keller (Eds.): Self-Managing Systems. 14th IFIP/IEEE International Workshop on Distributed Systems: Operations and Management (DSOM 2003). Heidelberg. Germany. (October 2003). Springer-Verlag. Vol. 2867. pp. 246–259

9. CONTEXT: Active Creation, Delivery and Management of efficient Context Aware Services. IST-2001-38142- CONTEXT. http://context.upc.es

10. Razzaque, M., Dobson, S., Nixon, P.: Categorisation and Modelling of Quality in Context Information. In Proceedings of the IJCAI 2005 Workshop on AI and Autonomic Communications. Edinburgh. Scotland. (August 2005).

11. Pipino, L., Lee, Y., Wang, R.: Data Quality Assessment. Communications of the ACM (April 2002). Vol. 45, No. 4. pp. 211-218

12. Cohen, N., Castro, P., Misra, A.: Descriptive Naming of Context Data Providers. In Proceedings of Fifth International and Interdisciplinary Conference on Modeling and Using Context (CONTEXT 2005). Paris. France. (March 2005).

13. Kellerer, H., Pferschy, U., Pisinger, D.: Knapsack Problems. Springer. ISBN: 3-540-40286-1. p. 546.(2004)

14. Hifi, M., Micrafy, M., Sbihi, A.: Heuristic Algorithms for the Multiple-choice Multidimensional Knapsack Problem. Journal of the Operational Research Society. Vol. 55. pp:1323-1332. (December 2004)

15. Akbar, M., Rahman, S., Kaykobad, M, Manning, E., Shoja, G.: Solving the Multidimensional Multiple-choice Knapsack Problem by Constructing Convex Hulls. To appear in Computers & Operations Research. Elsevier Science

16. Khan, S.: Quality Adaptation in a Multi-session Adaptive Multimedia System: Model, Algorithms and Architecture. PhD Thesis. Department of Electronical and Computer Engineering. University of Victoria. Canada. (1998)

17. Khan, S. Li, KF., Manning, E., Akbar, M.: Solving the Knapsack Problem for Adaptive Multimedia Systems. Studia Informatica. Special Issue on Combinatorial Problems. 2002. Vol. 2. No. 1. ISBN 2-912590-13-2. ISSN Regular 1625-7525.

Identifying the Multiple Contexts of a Situation

Aviv Segev

Technion - Israel Institute of Technology, Haifa 32000, Israel
asegev@technion.ac.il

Abstract. The paper presents a contexts recognition algorithm that uses the Internet as a knowledge base to extract the multiple contexts of a given situation, based on the streaming in text format of information representing the situation. Context is represented here as any descriptor most commonly selected by a set of subjects to describe a given situation. Multiple contexts are matched with the situation. The algorithm yields consistently good results and the comparison of the algorithm results with the results of people showed that there was no significant difference in the determination of context. The algorithm is currently being implemented in different fields and in multilingual environments.

Keywords: Matching context, Context recognition, Metadata, Text analysis.

1 Introduction

The question of context recognition is defined as one of the main questions addressed by the international interdisciplinary context community [1]. A context is a descriptor (such as a word or an image) or a set of descriptors that defines a situation. A context can convey a different facet, a different point of view, or a different understanding of a situation. Therefore, many situations are characterized by several different contexts. This paper presents an algorithm of contexts recognition that analyzes a given situation, represented in text format, and identifies its multiple contexts, textual descriptors. The performance of the algorithm was compared to the human process of multiple contexts recognition and yielded consistently good results. The algorithm is now being implemented in different fields and in a multilingual eGovernment project.

Section 2 reviews related works in the literature. Section 3 presents a formal definition of contexts recognition and divides the problem into two sub-problems. Section 4 describes the contexts recognition algorithm, which consists of five main processes: the collection of data, the selection of contexts for each text, the ranking of the contexts, the identification of the current contexts, and the clustering to achieve the multiple contexts, and describes the use of the Internet as a context database. Section 5 presents the analysis of the algorithm. Section 6 discusses the applications of the algorithm in different domains of knowledge and in different languages. Finally, section 7 presents some conclusions and directions of future research.

2 Related Work

2.1 Formal Definition of Context

Context is defined as a descriptor (such as a word or an image) or set of descriptors that can represent a situation or a scenario. A scenario is defined as "the world state", a

T.R. Roth-Berghofer, S. Schulz, and D.B. Leake (Eds.): MRC 2005, LNAI 3946, pp. 118–133, 2006.

situation that is a snapshot or an instance of the world at some given time, namely, all attributes of the world, including all objects, their properties and internal states, and the relationships between them [21].

McCarthy [18] defined the formalization of the notion of context as one of the main problems in the field of artificial intelligence (AI) and argued that a most general context does not exist. Consequently, McCarthy [19] worked to formalize context and to develop a theory of introducing context as formal objects.

Since a situation is characterized by many different features, it may have multiple contexts. McCarthy's formal definition of context is used in this work to identify the multiple contexts of a situation.

2.2 Blackboard Model

The model architecture for the contexts recognition is based on the blackboard model [7]. The blackboard model has been used in many AI applications, e.g., understanding images [24], signals [9], and speech [16], to represent possible solutions to a given problem. Blackboard is used here to enhance information extraction from more than one information source. For example, different information sources can be multiple people having multiple conversations at the same place and time, as in the case of Internet chats.

2.3 Context Extraction

Since virtually every application requires the use of context, whether explicitly or implicitly, it is necessary to have means by which to extract it. Bauer and Leake [4] developed WordSieve, an algorithm for automatically extracting information about the context in which documents are consulted during web browsing. Using information extracted from the stream of documents consulted by the user, the WordSieve algorithm automatically builds context profiles that differentiate sets of documents that users tend to access in groups. These profiles are used in a research-aiding system to index documents consulted in the current context and pro-actively suggest them to users in similar future contexts.

Another approach created taxonomies from metadata (in XML/RDF) containing descriptions of learning resources [20]. After the application of basic text normalization techniques, an index was built, observed as a graph with learning resources as nodes connected by arcs labeled by the index words common to their metadata files. A cluster mining algorithm is applied to this graph and then the controlled vocabulary is selected statistically. However, a manual effort is necessary to organize the resulting clusters into hierarchies. When dealing with medium-sized corpora (a few hundred thousand words), the terminological network is too vast for manual analysis, and it is necessary to use data analysis tools for processing. Therefore, Assadi [2] employed a clustering tool that utilizes specialized data analysis functions and clustered the terms in a terminological network to reduce its complexity. These clusters are then manually processed by a domain expert to either edit them or reject them.

We propose the use of a fully automatic contexts recognition algorithm that uses the Internet as a knowledge base and as a basis for clustering.

2.4   Information Seeking

Information seeking is the process in which people turn to information resources to increase their level of knowledge regarding their goals [8]. Although the basic concept of information seeking remains unchanged, the growing need for the automation of the process has called for innovative tools to assign some of the tasks involved in information seeking to the machine level. Thus, techniques for information seeking based on textual information are used, including the ontology tools Text-To-Onto [17], OntoMiner [11], and TexaMiner [12], to name a few, and databases are extensively used for the efficient storage and retrieval of information.

The Internet can be seen as a large database that is constantly being modified and updated. Many information seeking techniques have been developed to retrieve information from the Internet. For example, Valdes-Perez and Pereira [23] developed an algorithm based on the concise all pairs profiling (CAPP) clustering method. This method approximates profiling of large classifications. The use of hierarchical structure was explored for classifying a large, heterogeneous collection of web content [6]. Another method involves checking the frequency of the possible keyphrases of articles using the Internet [22]. However, this method is based on an existing set of keywords and uses the Internet for ranking purposes only.

The present algorithm attempts to automate contexts recognition, based on information seeking techniques, using the Internet as a database for possible multiple contexts. The algorithm differs from previous text analysis techniques by allowing the input to be received from multiple sources, in an unstructured format. In addition, the algorithm utilizes data resources that are independent of the user and are constantly changing to analyze the information.

3   Formal Definition of the Problem

A scenario can be characterized by multiple contexts, each describing a different facet of the situation.

McCarthy [19] formalized context as first class objects with the following basic relation:

ist(C, P) meaning that the proposition P is true in the context C.

In this paper, context is defined as any textual description that is most commonly selected by a set of subjects to describe a given situation and multiple contexts are a set of such contexts:

Let $P_1, P_2, ..., P_m$ be a series of textual propositions defining situation S.

Contexts $C_1, C_2, ..., C_k$ are defined as the contexts of situation S if:

$\exists n$ subjects, $n \geqslant 1$ so for the majority of n selected

ist(C_i, P_j) $\forall$ i, for a given j

(Contexts $C_1, C_2, ..., C_k$ are true for textual proposition P_j)

For a series of propositions there exists a collection of sets of contexts.

Let $P_1, P_2, ..., P_m$ be a series of textual propositions when $\forall$ P_i there exists a collection of sets of contexts C_{ij} so that:

$\forall$ i, ist($\mathcal{C}_{ij}$, P_i) $\forall$ j meaning that the textual proposition P_i is true in each of the set of contexts $\mathcal{C}_{ij}$. $\mathcal{C}_{ij}$ are not predefined hierarchically in a structure such as a tree. However, hierarchical structures can be built according to a specific set of textual propositions.

The main research problem is formally defined as:

What is the outer context $\mathcal{C}$, the multiple contexts of a scenario, defined by

$ist(\mathcal{C}, \bigcap_{i=1}^{m} ist(\mathcal{C}_{ij}, P_i))$ $\forall j$.

The number of existing contexts is assumed to be finite and to satisfy

$\mathcal{C}, \mathcal{C}_{ij} \subseteq U_c$ (unity of all existing contexts)

The main research problem can be divided into two sub-problems:

1. Let P be a given single text. What are the possible contexts $\mathcal{C}_i$ that satisfy ist($\mathcal{C}_i$, P) $\forall$ i (for single text P all contexts $\mathcal{C}_i$ are true)
2. Let $P_1, P_2, ..., P_m$ be a set of texts that satisfy the following condition: for each text P_i there exists a set of contexts $\mathcal{C}_{ij}$ so that ist($\mathcal{C}_{ij}$, P_i) $\forall$ j.
   What is the outer context $\mathcal{C}$ so that $ist(\mathcal{C}, \bigcap_{i=1}^{m} ist(\mathcal{C}_{ij}, P_i))$ $\forall j$.

The division of the problem into two parts allows a solution of the first part to be acquired by information seeking through the Internet. The second part is addressed using an algorithm that ranks the contexts according to the importance of the information retrieved in the first part. The result of the following algorithm is a list of contexts, the outer context of the situation, which is the multiple contexts of the situation.

4   The Contexts Recognition Algorithm

The research algorithm is based on the streaming in text format of information that represents input from different sources, such as Internet chats. The contexts recognition algorithm output is a set of contexts that attempt to describe the current situation most accurately. The set of contexts is a list of words or phrases, each describing an aspect of the situation. Thus, multiple contexts can be matched to a given situation. The algorithm consists of five major processes:

- Collecting Data - The information from the information sources is decomposed into words and the keywords are extracted from them.
- Selecting Contexts for Each Text (Descriptors) - For each keyword a set of preliminary contexts is extracted from the Internet, which is used as a context database.
- Ranking the Contexts - Each preliminary context is ranked according to the number of references it receives in the context database and the number of appearances it has in the text.
- Identifying the Current Contexts - The preliminary contexts that have significantly higher numbers of references and higher numbers of appearances are included in the current set of contexts.
- Obtaining the Multiple Contexts - The current contexts are examined for synonyms and synonymous contexts are united.

4.1   Collecting Data

The input text was used as is; all misspelled words were left in the text. The text was parsed at the granularity of sentences. Long sentences were parsed according to the maximum number of words that could be used in a search engine. Each text is decomposed into single words, when words are letter strings separated by spaces, and all punctuation is removed from the text. Then the words are checked according to a set of dictionaries. The first dictionary is a "Stop List", consisting of words that do not add to the understanding of the context, such as I, me, in, are, the. All words that appear in this dictionary are ignored. The next step uses a set of dictionaries according to fields of knowledge to sieve the words that are not related to the specific field of knowledge. If the word appears in the field of knowledge dictionary, then it is added to the list of keywords that are searched in the context database, otherwise it is ignored. This process continues for each word in the text. After each text passes through this module, the algorithm sends a list of words to be checked for a possible set of contexts. Checking against a dictionary can be skipped if the field of knowledge is unknown, but skipping this step may sometimes lead to less accurate results. The difficulty does not lie in finding the possible keywords since the algorithm can always use the whole input corpus.

Algorithm 4.1: COLLECTING DATA($TextualData$)

Parse data according to the granularity of sentences
Replace punctuation with a new line
Eliminate words which appear in "Stop List"
if Field of Knowledge is defined
   then Eliminate words that do not appear in field on knowledge dictionary
if words in line $>$ maximum words for search engine
   then Create new lines according to maximum words
for each new line
   do Activate the next algorithm

4.2   Selecting Contexts for Each Text (Descriptors)

The selection of the current context is based on a search through the database for all relevant documents according to keywords and on the clustering of the results into possible contexts. Once a list of keywords exists, each keyword is searched in the context database - the Internet - and a set of contexts is extracted. This creates a list of preliminary contexts for each keyword. The contexts in this work were represented by words or sets of words, which can be viewed as meta data created for each set of Internet web pages. The Internet can then be viewed as an immense set of words that represent different possible contexts, each associated with its respective web page. Other descriptors can include images appearing on the Internet. The Internet can then be seen as a vast set of descriptors that represent different possible contexts, each associated with its respective web page. The full list of preliminary contexts for all the keywords includes all the possible contexts for this current text.

Any search engine can be used and any Term Frequency / Inverse Document Frequency [10] method for clustering can be implemented. The current application of the algorithm uses the concise all pairs profiling (CAPP) clustering method. [23], as it is applied in the Vivisimo search engine.

The use of the Internet as a context database instead of a precalculated frequencies base has several advantages. The use of the Internet does not require the constant updating and maintenance of a database. The precalculated frequencies base requires the user to work in a limited predefined knowledge domain. The Internet can serve as an unlimited knowledge domain that is continuously being updated.

This step results in a long list of preliminary contexts, many of which are irrelevant to the context. The purpose of the following steps is to minimize the list and identify most relevant contexts of the situation.

Algorithm 4.2: SELECTING CONTEXTS($List of Words$)

Check Internet search engine with *List of Words*
for each Internet page extracted
  do Activate Ranking Contexts algorithm
Add 1 to *Number of Appearances* for each context identified

4.3   Ranking the Contexts

The algorithm checks the number of appearances in the text for each preliminary context in the set of preliminary contexts. The contexts are also examined for the number of Internet documents that refer to the set of documents. The set of contexts is now ranked according to both the number of references in the text and the number of references in the documents.

These two metrics were selected since the number of appearances in the text represents how many times each preliminary context was mentioned in the situation. The number of references in the Internet represents how important the preliminary context is to the general population that uses the Internet.

New preliminary contexts can now be created according to textual sub-strings of existing preliminary contexts. This step sums up the number of documents referring to the preliminary contexts. Multiple reference pages from similar web sources are counted as one instance. Each document usually refers to multiple contexts, consequently creating a long list of preliminary contexts. The last step involves ranking the set of preliminary contexts according to both the number of references from the documents and the number of appearances in the text. This step maps all the preliminary contexts to a two dimensional graph, allowing the contexts that receive very high ranking in both characterizations to be located, as in Figure 1.

After each session of ranking, the list is used for two purposes - resetting the set of preliminary contexts and identifying the current context. The current list of contexts joins the new preliminary contexts arriving from the continuously streaming text. The lists are united and the ranking process is repeated. In parallel to the repetition of the ranking algorithm, the set of ranked preliminary contexts is forwarded to the next module to determine the current contexts.

Algorithm 4.3: RANKING CONTEXTS($Internet Page Extracted$)

Perform term frequency clustering
for each term
  do $\begin{cases} \textbf{if } \text{term not in } \textit{Preliminary Context List} \\ \quad \textbf{then } \text{Add term to } \textit{Preliminary Context List} \end{cases}$

4.4 Identifying the Current Contexts

The output of the ranking stage is the current context or a set of highest ranking contexts that differ essentially by rank. The algorithm then returns to the first step to collect more texts and feed them again to the database. The set of preliminary contexts that has the top number of references, both in number of Internet pages and in number of appearances in all of the texts, is defined as the highest ranking and is identified to be the current contexts.

The current contexts received from the previous stage can be depicted on a graph according to the number of appearances and the number of references, as in Figure 1.

The algorithm for detecting the current contexts includes the following steps:

Algorithm 4.4: DETECTING CURRENT CONTEXTS($PreliminaryContexts$)

Organize the list of preliminary contexts in descending order according to number of references appearing in the Internet - the Set of Documents.

Find the difference between each value of the number of references and its nearest lower value neighbor, defined as Current References Difference Value (CRDV).

Find the difference between each value of the number of appearances and its nearest lower value neighbor, defined as Current Appearances Difference Value (CADV).

Weight the number of appearances in the text and the number of references in the Internet according to the following formula:
MVR = Maximum Value of References
MVA = Maximum Value of Appearances

$$Weighted Value = \sqrt{(\tfrac{2*CADV*MVR}{3*MVA})^2 + (CRDV)^2}$$

Find the maximum value of the Weighted Value. If the maximum Weighted Value is the first value, then continue to the next one, since frequently the first value is too far from its neighbor.

Select all the contexts that appear before the maximum Weighted Value in the list that was organized in the first step as the current contexts. Store current selected contexts.

Erase the selected contexts from the list and repeat the previous two steps.

The Weighted Value can be viewed as the weighted distance to the origin. However, the index of number of references is on a much larger scale than the index of the number of appearances and therefore it is not possible to retain the original proportions and it is necessary to re-scale the indices. The value is calculated multiplying by the maximum number of references and dividing by the number of maximum number of appearances. A constant of 2/3 was found by experiment to be appropriate for the re-adjustment of the figures.

The first cluster of contexts near the origin includes all the contexts that received low ranking both in number of appearances and in number of references. This group of contexts includes most of the contexts in the list. Since the contexts in this group received low ranking they are eliminated from the list. The remaining contexts are the current contexts.

The process can continue until all the contexts in the list are covered and this will yield all the possible preliminary contexts. However, in most cases the best results were already achieved when the last two steps were performed twice. Further repetitions, which increase the number of results, were unnecessary. The ranking according to number of references and not according to the weighted value also improves the results. This indicates that the number of appearances of the context in the text has less value in the determination of the context than the number of references in the Internet.

During the implementation of the algorithm, there was a problem that required special consideration. The contexts that received lower ranking than the top ranking contexts in the cluster but were not part of the cluster were kept. Namely, these are contexts that receive lower ranking in either the number of appearances or the number of references than the top ranking contexts, but not in both. Running the algorithm showed that these contexts are sometimes relevant and should be kept.

4.5   Obtaining the Multiple Contexts

The current contexts are examined for synonyms using a thesaurus and synonymous contexts are united. Before this step, many of the contexts identified by the algorithm are similar in meaning. The algorithm also looks for semantic similarity. This step enables the algorithm to identify the differing multiple contexts of the situation and thus facilitates the better description of the situation.

Algorithm 4.5: MULTIPLE CONTEXTS CLUSTERING($CurrentContexts$)

for each $x \in CurrentContexts$

do $\begin{cases} \text{Examine the } CurrentContexts \text{ for synonyms in Thesaurus} \\ \text{Examine the } CurrentContexts \text{ for semantic similarity} \\ \textbf{if } \text{synonyms / semantic similarity found} \\ \quad \textbf{then } \text{Unite contexts and unite their weights} \end{cases}$

return (United Contexts, Relevant Weights)

The current set of contexts is the output of the algorithm. However, since the algorithm is continuous, the contexts continue updating as long as new textual input continues to be accessed by the algorithm.

4.6   Examples

Example 1. The example presents the implementation of the algorithm on text taken from MSN chat and the results of the algorithm.

Lestat: Question on Linux how much ram can it run
WickedWeekend: like i said im new at this
Xor: ifconfig eth0 192.168.0.1 netmask 255.255.255.0
Xor: lestat virtually any amount of RAM
Xor: if u adjust kernel to it
WickedWeekend: ok Xor
Lestat: well you know my computer 1.53gb of ram
WickedWeekend: to me just looking at it and not being an expert
Xor: lestat it will suffice
Xor: wykd
Xor: yes there are many other command
Xor: but good thing about linux
Xor: is that u cna make aliases to commands
Xor: so instead of ifconfig eth0 192.168.0.1 netmask 255.255.255.0
Xor: u can make an alias
Xor: in form
WickedWeekend: if your ethernet configuration ip is 192.168.0.1 you want it to have a mask of 255.255.255.0 which is a generic one i think instead of broadcasting your native ip?
Xor: dothis
Xor: u cna have any subnet mask
Xor: u want
Xor: u can mak eur own subnetmasks
Xor: 255.255.248.0
Xor: or whatever
WickedWeekend: can they be virtualy anything as long as they are in the correct format?
WickedWeekend: but have to start 255.255.........
Xor: 255.255 not necesserily
Xor: u can have
Xor: 255.248.0.0
Xor: anythign really
WickedWeekend: but the fist one is always 255?
WickedWeekend: or not?
Xor: well yes it should be
Xor: i never saw another format
WickedWeekend: ok

First the input is read one line at a time. Each word is separated by a space. Punctuation marks are eliminated. Each word is checked against the "Stop List" dictionary. In this case each word was checked in a predefined computer dictionary.

The words that passed the previous stage serve as keywords. After each step (change of speaker) the keywords are sent to the search engine and clustered into a list of preliminary contexts. These steps are repeated 34 times, yielding 222 preliminary contexts that have at least two references in the Internet and are relevant to keywords that appeared at least once in the text.

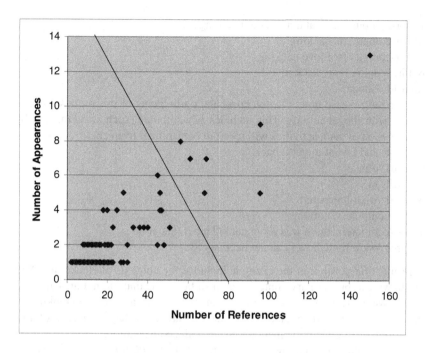

Fig. 1. Identifying the Current Contexts

The algorithm maps all the preliminary contexts to a two dimensional graph, allowing the contexts that receive very high ranking in both characterizations to be located, as in Figure 1.

The set of identified contexts includes: Linux(150,13), Subnet(96,5), Forums(96,9), Review(69,7), FAQ(68,5), Blog(61,7), and Network(56,8). The value in the parentheses includes the number of references and the number of appearances respectively. Figure 1 shows the weighted value calculated for each context. The differences between the weighted values in the figure show that the maximum weighted value is after Network, resulting in the current contexts.

The chat can be viewed as having multiple contexts. The contexts Network and Subnet represent the communication point of view discussed between two participants in the chat. At the same time, a discussion about Linux is being held. Looking from a broader perspective, the contexts of Forums and Frequently Asked Questions (FAQ) can be viewed as the contexts of the conversation.

Example 2. The example presents the implementation of the algorithm on text taken from MSN chat and the results of the algorithm. The algorithm was run without any information about the knowledge domain and did not use any field of knowledge dictionary.

Fox-Fire1: i want to ask some question about hacking
Fox-Fire1: any body help me
simply-crazy: h word is bad

mad-for-computers: what ques about haciking

mad-for-computers: hacking

mad-for-computers: i love hacking

Fox-Fire1: some body hack my id

Demon11: !illegal

Obis-Shadow: Please note: Owner, hosts and participants of this room we don't offer any help with illegal activity. This includes p2p software.(such as Kazaa and Imesh). any discussion of such activities will result in banishment from chat.

Fox-Fire1: and i want to get it back

Fox-Fire1: ohh

Fox-Fire1: hi

Fox-Fire1: what happened

mad-for-computers: nothing

mad-for-computers: there was room named h but it is gone

Fox-Fire1: o k tell me how i can get back my id

The results of algorithm in the example included the contexts of Hacker and Security. This shows that preliminary contexts can also be words that did not appear in the text itself but were a result of the clustering of the web pages, as in the top ranking context of Security. The contexts of Hacker and Security are the different contexts of this chat, representing different complementary facets of the conversation.

Demon11 and Obis-Shadow are software bots that monitor the chat. Eventually the participant Fox-Fire1 was banished from the chat as a result of raising an illegal discussion topic - hacking, although the participants were discussing security measures against hacking.

5 Analysis

The objective was to analyze the different multiple perspectives, or contexts, of a given scenario, with minimal restrictions placed on the human subjects evaluating the scenario so as not to direct them to a single perspective. The contexts recognition algorithm was evaluated using computer-related Internet chats acquired from MSN chats. The chats included several participants and were observed over time. Parts of the chats that dealt with topics concerning computers were copied to files. From these chats sets of files were randomly selected to be analyzed by the algorithm. These files were fed as input into the contexts recognition algorithm. The results were compared with the results given by computer-literate subjects.

The subjects who answered the survey were graduate students with at least basic knowledge of computer terminology. A total of twenty subjects participated in the survey. The participants received a set of three chats. This allowed two groups of participants to be formed, each group with a different set of chats. Each chat had at least nine replies. The maximum number of replies per chat was eleven. An average of ten subjects determined the list of contexts for each chat. A total of six chats in computer related topics were analyzed in this way. The subjects were presented with the above chats and were asked to provide a list of contexts for the chats. The subjects were told that the text was obtained from Internet chats and was presented to them as is, including spelling mistakes and Internet acronyms. The subjects were asked to write in their own words what they felt were the contexts. These words were not selected from a list. The words were counted to determine the number of times they appeared among all the subjects. Each word was counted once for each subject who mentioned the context. The context was ranked in descending order according to the number of times that the subjects mentioned the context. The list of best-ranked contexts was compared with the results yielded by the contexts recognition algorithm. The other contexts mentioned more than once by the subjects were also compared to check the sensitivity of the algorithm.

Table 1 summarizes the comparison of the results of the algorithm and the results of the subjects. The table also displays the average results of the contexts recognition algorithm for all six cases examined. The second and third sets of contexts, which are contexts mentioned by the subjects but not selected by a majority of the people or only selected by one or two subjects, respectively, are also compared with the second and third reiterations of the algorithm to check the sensitivity of the algorithm.

Table 1. Ranked Contexts (RC) of the Algorithm

Contexts Recognition	Chat 1	Chat 2	Chat 3	Chat 4	Chat 5	Chat 6	Average
Top RC	100%	100%	100%	100%	100%	100%	100%
Top and II RC	100%	84.46%	65.22%	57.15%	75%	100%	80.31%
Top, II, and III RC	86.67%	81.25%	57.15%	40%	76.92%	75%	69.50%

The top ranking contexts mentioned by the subjects were identified as contexts by the algorithm. In most of the cases the contexts that were ranked among the highest by the subjects were also ranked among the highest by the algorithm. Some of the other contexts generated by the algorithm were not selected by the subjects. In addition, a few of the lower ranking contexts mentioned by the subjects were missed by the algorithm. Table 1 shows that for the top ranking contexts the algorithm yields very high results. As more contexts that received lower ranking by the subjects are added, the results of the algorithm degenerate. The algorithm needs improvement in deducting better results in the second and third ranking sets of contexts.

The significance of the results was analyzed using the identical populations test. The test for homogeneity is designed to test the null hypothesis that two or more random samples are drawn from the same population or from different populations, according to some criterion of classification applied to the samples. The Chi-square Pearson Test for Association is a test of statistical significance. The results of the identical populations test comparing the groups containing the algorithm as a subject with the original group consisting only of human subjects showed that they were almost identical populations. In other words, if the computer is part of the group, the context will remain identical. Hence, there is no significant difference in the determination of contexts between the algorithm and the human subjects. Table 2 displays the identical population test results for each of the six chats.

The complexity of the algorithm is $\theta(kn)$ where n represents the number of input cycles such as each line of text or each time that input is received from a different source. The k represents a constant limiting the number of top ranking results from each

Table 2. Identical Populations Test

Chat	χ^2	P-Value
1	0.065199	1
2	0.568693	0.999
3	0.187391	1
4	0.256795	1
5	0.133712	1
6	0.273300	0.998

cycle of the algorithm. This allows different levels for the monitoring of the amount of data the algorithm handles.

6   Applications

6.1   Applicability in Multiple Domains

The contexts recognition algorithm is versatile in terms of its utility in multiple domains of knowledge. The algorithm was extensively analyzed using Internet chats on a wide variety of topics, including health-oriented chats and computer-oriented chats. The algorithm yielded consistently good results in this broad range of topics.

The algorithm can be used without the pre-definition of the field of knowledge. Currently the algorithm is being implemented in the field of medical case studies, with the use of a field-specific dictionary, and in the field of eGovernment services, without any such pre-definition.

In the field of medical case studies, the contexts recognition algorithm is being used to extract information from actual medical cases. The goal is to examine a method for encapsulating a patient's medical history and current situation into keywords - the contexts of the medical case studies - so as to assist the physician in his analysis.

In the field of eGovernment services, the algorithm is currently being examined in TERREGOV and QUALEG, European IST projects. TERREGOV aims at providing territorial governments with flexible and interoperable tools to support the change towards eGovernment services. The purpose is to identify the contexts of documents to enable the revision of ontologies for the optimization of the indexing and search of documents. QUALEG aims at providing local governments with an effective tool for bi-directional communication with citizens. Contexts are used to specify citizen input and then provide services - routing emails to departments, opinion analysis on topics at the forefront of public debates, and identification of new topics on the public agenda.

6.2   Applicability in Multilingual Settings

The contexts recognition algorithm is also versatile in regards to the language of the input text. The algorithm enables the identification and representation of the context in multiple languages. The algorithm is not language dependent, since the Knowledge Base is extracted from the Internet. The algorithm success rate depends on the number of Internet pages existing in each language.

The Web is a multilingual corpus. Xu [25] estimated that 71% of the pages (453 million out of 634 million Web pages indexed by the Excite search engine at that time) were written in English, followed by Japanese (6.8%), German (5.1%), French (1.8%), Chinese (1.5%), Spanish (1.1%), Italian (0.9%), and Swedish (0.7%).

One hundred million words is a large enough corpus for many empirical strategies for learning about language, either for linguists [3] and lexicographers [14] or for technologies that need quantitative information about the behavior of words as input (most notably parsers [5][15]). However, for some purposes, it is not large enough.

Our initial experiments in the QUALEG project show results that coincide with the above data. The previous section displayed the consistently good results of the algorithm in English (See Table 1). Analysis of email contexts yields a high success rate of 84% in the German language as well. However, for the Polish language which has 0.42% of the web pages in the English language [13] the success rate of the algorithm is much lower and thus complementary techniques from Natural Language Processing are currently being integrated to increase effectiveness.

7   Conclusion

Every situation can be characterized by multiple contexts that describe its different aspects and that are necessary for the complete understanding of the different perspectives of the situation. The main idea of the research was to use the Internet as a contexts database to identify the multiple contexts of the given situation. The Internet is a source of information that is constantly increasing and being updated. The use of the Internet as a database for contexts recognition therefore gives a contexts recognition model immediate access to a nearly infinite amount of data in a multiplicity of fields. Hence, the necessity of creating a database for the determination of the contexts is eliminated.

Furthermore, the situations for which the contexts are sought can be independent of the Internet; the Internet is merely the database in which the algorithm searches for the contexts. Thus, for example, the contexts of a conversation between people can be found through the use of the Internet - the algorithm is a tool that allows the computer to determine the contexts by using the Internet as a database and then to pass these contexts back into the real world. The Internet is one possible source of data, but the algorithm holds also for a more restricted database. Intranet data, internally generated textual information about the organization that is stored, can also be used.

Another advantage of the contexts recognition algorithm is that it functions in real-time without needing a period of training or practice. Thus, it extracts the contexts immediately with little previous user intervention.

Tests show that the algorithm also achieves good contexts recognition results without the use of a field of knowledge dictionary, which represents specialized knowledge. Thus the algorithm can be used in diverse areas without predefined knowledge of the field.

The complexity of the algorithm is directly dependent on the size of the input description of a given situation. Thus, online implementation is feasible. Moreover, the algorithm can be implemented in an extensive variety of domains, since it is field independent. Current implementations of the algorithm focus on medical case studies and online eGovernment applications. These online eGovernment applications show that the algorithm is also language independent and can be implemented in multilingual settings.

The Internet includes many different representations of data, such as text, image, and sound. Therefore, future directions of research include implementing the algorithm to extract contexts in alternative formats of representation. Other directions of research include mapping multiple contexts to ontologies, since contexts and ontologies are complementary disciplines of modeling views.

Acknowledgments

The work of Segev was partially supported by two European Commission 6th Framework IST projects, TerreGov (http://www.terregov.eupm.net) and QUALEG (http://www.qualeg.eupm.net), and the Fund for the Promotion of Research at the Technion.

References

1. In *AAAI'99 (1999) Workshop on Reasoning in Context for AI Applications*, July 19 1999.
2. H. Assadi. Construction of a regional ontology from text and its use within a documentary system. *In Proceedings of the International Conference on Formal Ontology and Information Systems (FOIS-98)*, 1998.
3. C. F. Baker, C. F. Fillmore, and J. B. Lowe. The berkeley framenet project. In *Proceedings of COLING-ACL*, pages 86–90, 1998.
4. T. Bauer and D. Leake. Wordsieve: A method for real-time context extraction. In *CONTEXT 2001*, pages 30–41, 2001.
5. T. Briscoe and J. Carroll. Automatic extraction of subcategorization from corpora. In *Proceedings of the Fifth Conference on Applied Natural Language Processing*, 1997.
6. S. Dumais and H. Chen. Hierarchical classification of web content. In *Proceedings of SIGIR, 23rd ACM International Conference on Research and Development in Information Retrieval*, pages 256–263, 2000.
7. L. Erman, F. Hayes-Roth, V. Lesser, and D. R. Reddy. The Hearsay II speech understanding system: Integrating knowledge to resolve uncertainty. *Computing Surveys*, 12(2):213–253, 1980.
8. A. Gal, G. Modica, H.M. Jamil, and A. Eyal. Automatic ontology matching using application semantics. *AI Magazine*, 26(1), 2005.
9. N. Gerard and V. Lesser. *Blackboard Systems for Knowledge-Based Signal Understanding, Symbolic and Knowledge-Based Signal Processing*. Prentice Hall, 1992.
10. S. Gerard. *Automatic Text Processing: The Transformation, Analysis, and Retrieval of Information by a Computer*. Addison-Wesley Publishing Company, Inc., 1989.
11. S. Vadrevu H. Davulcu and S. Nagarajan. Ontominer: Bootstrapping and populating ontologies from domain specific websites. In *Proceedings of the First International Workshop on Semantic Web and Databases*, 2003.
12. V. Kashyap, C. Ramakrishnan, C. Thomas, and A. Sheth. Taxaminer: An experimentation framework for automated taxonomy bootstrapping. *International Journal of Web and Grid Services, Special Issue on Semantic Web and Mining Reasoning*, September 2005. to appear.
13. A. Kilgarriff and G. Grefenstette. Introduction to the special issue on the web as corpus. *Computational Linguistics*, 29(3), 2003.

14. A. Kilgarriff and M. Rundell. Lexical profiling software and its lexicographical applicationsa case study. In *Proceedings of EURALEX 02*, 2002.
15. A. Korhonen. Using semantically motivated estimates to help subcategorization acquisition. In *Proceedings of the Joint Conference on Empirical Methods in NLP and Very Large Corpora*, pages 216–223, 2000.
16. V. Lesser, R. Fennell, L. Erman, and D. R. Reddy. Organization of the hearsay ii speech understanding system. *IEEE Transactions on Acoustics, Speech, and Signal Processing*, 23:11–24, 1975.
17. A. Maedche and S. Staab. Ontology learning for the semantic web. *IEEE Intelligent Systems*, 16, 2001.
18. J. McCarthy. Generality in artificial intelligence. *Communication of ACM*, 30:1030–1035, 1987.
19. J. McCarthy. Notes on formalizing context. *In Proceedings of the Thirteenth International Joint Conference on Artificial Intelligence*, 1993.
20. C. Papatheodorou, A. Vassiliou, and B. Simon. Discovery of ontologies for learning resources using word-based clustering. *Proceedings of the World Conference on Educational Multimedia, Hypermedia and Telecommunications (ED-MEDIA 2002)*, pages 1523–1528, 2002.
21. R. M. Turner. Model of explicit context representation and use for intelligent agents. In *1999 International and Interdisciplinary Conference on Modeling and Using Context (CONTEXT-99)*, 1999.
22. D. P. Turney. Mining the web for lexical knowledge to improve keyphrase extraction: Learning from labeled and unlabeled data. ERB-1096 NRC #44947, National Research Council, Institute for Information Technology, 2002.
23. R. E. Valdes-Perez and F. Pereira. Concise, intelligible, and approximate profiling of multiple classes. *International Journal of Human-Computer Studies*, pages 411–436, 2000.
24. T. Williams, J. Lowrance, A. Hanson, and E. Riseman. Model-building in the visions system. In *Proceedings of IJCAI-77*, 1977.
25. J. L. Xu. Multilingual search on the world wide web. In *Proceedings of the Hawaii International Conference on System Science (HICSS-33)*, 2000.

An Engineering Approach to Adaptation and Calibration

Michael Fahrmair, Wassiou Sitou, and Bernd Spanfelner

Technische Universität München, Department of Informatics,
Boltzmannstr.3, D-85748 Garching (Munich), Germany
fahrmair@in.tum.de, sitou@in.tum.de, spanfeln@in.tum.de

Abstract. A new computing era after Mainframes, PC's and mobiles is becoming closer to reality since the beginning of the 21st century. This new era is often described with different terms such as pervasive, ubiquitous, ambient or context-aware computing. However, there is a common characteristic behind all these projections: They are all based on a substantially more flexible system understanding, whereby the thought of the system as a tool moves into the background and the needs and desires of the user step into the foreground. Such concepts for software applications being aware of their context are in fact not new, but become more and more important for productive fields of software and systems engineering and particularly in ubiquitous and wearable computing. In this paper we describe a generic mechanism for designing context awareness and adaptation behavior with formal methods, thus basically allowing an engineering approach in designing and implementing complex context aware adaptive systems while avoiding their usual pitfalls.

1 Introduction

Adaptation in a common sense is defined as an act of changing (structure, form, or habits) to fit different environmental conditions [13]. For technical systems in general such environmental conditions are usually referred to as *context* (see also [5] and [11]).

Nowadays, adaptation needs to be considered as a key requirement for future mobile and ubiquitous systems that envision heterogeneous environments where system and application functionality needs to be dynamically adapted to constantly changing situations like roaming access across different device capabilities and user personalization needs. There are existing architectures and frameworks such as [3] that more or less support developing context aware software. However, the important aspect of designing contexts and adaptation logic itself is typically overlooked. Moreover the context model and adaptation decision logic are usually static and hard coded in the adaptable entities. In view of ubiquitous systems, this approach seems inadequate, since the circumstances in which a system's functionality may be executed and the context parameters that may influence it, will not always be predictable a priori at the time the function is being developed. Generic, reusable mechanisms that offer runtime customizable context criteria and adaptation algorithms can offer a solution to this problem.

T.R. Roth-Berghofer, S. Schulz, and D.B. Leake (Eds.): MRC 2005, LNAI 3946, pp. 134–147, 2006.

In this paper we sketch out the scope of adaptation, describing what to be understood by adaptation and why this concept needs a particular consideration in software engineering research. After this we introduce a mathematical founded approach for designing flexible multiple (manual, automatical or combined) adaptations.

2   The Scope of Adaptation

The main goal of adaptation is to achieve *ubiquity*. Ubiquity means enhancing usability of functionality in as many situations as possible. A typical example for this is enabling the user to maintain a certain application while roaming between different radio access technologies, locations, devices and even simultaneously executing everyday tasks like meetings, driving a car etc. It is therefore important to consider the main ubiquity criteria, while dealing with adaptation.

In the following, we will start with a brief description of these criteria and of course the two main problems emerging from them. Both problems, the availability and the usability of adaptive applications and/or functionalities, need to be addressed. We propose in section 2.2 the approaches of reconfiguration and adaptation to address the availability and the usability problems respectively. The CAWAR context model will be described in section 2.3 including the necessary consideration for sensors, interpreters, actuators and context elements. At the end of this chapter, the overall process of adaptation will be presented, along with the logical architecture and the formalization of adaptation.

2.1   Ubiquity Criteria and Emerging Problems

For a certain functionality to be ubiquitous, three essential conditions need to be met:

- The necessary HW/SW infrastructure can be made available for the functionality in the given situation (*availability*),
- the functionality can fulfill the current user requirements (*applicability*) and
- the necessary interactions are not conflicting with the user's situation, this means range within his current free interaction possibilities (*operability*).

Looking into systems that should provide (implement) a ubiquitous functionality, it is obvious that a maximum ubiquity cannot be achieved by stacking up implementations for static requirements as is done usually in nowadays nonadaptive systems. A principal reason for this fact is that the number of requirements implemented into a system is limited by available resources (CPU, memory, bandwidth etc.). Moreover these resource limitations can vary from situation to situation (battery power or available networks). The resource limit is therefore usually given by tradeoffs between the least common denominator of providable functionality and a restriction of (availability-) situations resulting in reduced ubiquity. This problem reflects the availability dimension of ubiquity and therefore is called *availability problem*.

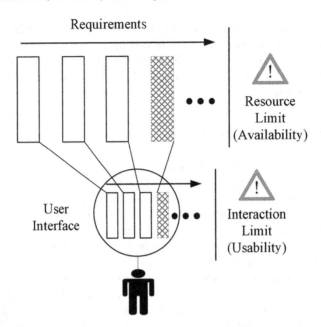

Fig. 1. Requirements Superposition and Limited Ubiquity (Availability and Usability)

Besides availability, there is yet another problem regarding nonadaptive systems that prevents ubiquity. Stacking up implementations of predefined requirements into a superposition of all requirements a system needs to fulfill during its life-time generates additional interactions between a system and its user, besides the interactions necessary to use the actual functionality. These additional interactions are necessary to communicate the current user's requirements selection in that given situation, so that they can be mapped onto the superposition of requirements fulfilled by the system. The amount and complexity of interactions between a user and a system is, however, also often limited by the current situation. For example while it might be possible for a user to navigate through a complicated menu structure while sitting in a train, the same interaction is not conceivable for him while driving a car or using another functionality that already requires his exclusive attention (e.g. a groupware conference). Without adaptation, there are obvious tradeoffs between type, amount and complexity of user interactions and the number of functionalities a system can provide; again limiting the number of (user-) situations, where a given functionality can be used. Analogue to availability this reflects the usability dimension of ubiquity, and therefore is called *usability problem*. Figure 1 illustrates both the availability and the usability problems.

2.2   Reconfiguration and Adaptation

Introducing *reconfiguration* solves the availability problem. This is due to the fact that a superposition of static requirements can be clustered into groups of non conflicting requirements that are usually required by the user in a certain situation. Such a cluster of implemented requirements is called *configuration*. A system is *reconfigurable* if it supports changing between different sets of such configurations according to the momentary situation. While it is possible to have reconfigurable systems that are managed by the users (e.g. a PC to install and choose between several applications), this kind of solution does not necessarily add to ubiquity. This is because reconfiguration itself adds additional user interactions in deciding about and choosing the proper configuration for a certain situation.

To solve the usability problem of ubiquity, another mechanism besides reconfigurability needs to be introduced. This mechanism should automate the reconfiguration management on behalf of the user (or other person entities with influence on requirements that should be fulfilled by a certain configuration). To achieve this, all or part of the manual reconfiguration management can be replaced with further functionalities fulfilling additional requirements of automatically reconfiguring between dynamic requirement implementations in different situations [16].

2.3   The CAWAR Context Model

To enable reconfiguration management functionality, the system needs to decide on behalf of the user entity about its current requirements. This can be done by monitoring the system's environment, including all users and the internal state of the system itself [23]. Environment monitoring is done using a context. A *context* is the sufficiently exact characterization of the situations of a system by means of perceivable information that is relevant for the adaptation of the system. This means, the context is a model of a situation that contains all necessary and available information to reason about the user's (or any other stakeholder involved in the system's usage) requirements. Since context is usually defined using an abstract view of a situation (a model) [12], a specific context is always specified from a certain perspective and describes only the relevant information from the system's environment. For example a user profile [8] could be used as a context describing user preferences [20].

A simple context can be modeled using an entity relationship data model that holds the contextual information. The model proposed here is more detailed and differentiates in sensors, context data and interpreters as proposed in [5], to enrich the static data structure of context with its dynamic processing information. *Sensors* observe the external system environment. They gather information that describes the system's situations and their changes, and update the sensor context data accordingly. Intermediate context data in contrast is updated by interpreters observing the sensor context. *Interpreters* can calculate any information that can not directly be measured with sensors thus forming an abstract (interpreted) description of the initial physical (sensored) situation. A change of an intermediate context can of course trigger other interpreters, resulting in further context data changes and so on. Since such a context model does not only describe contextual information but also its sources, it has the advantage

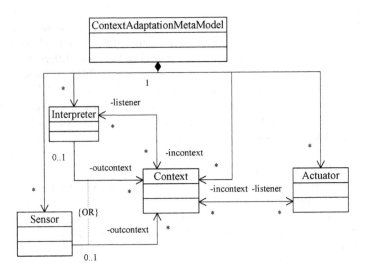

Fig. 2. Context Adaptation Metamodel

that new sensors and interpreters can be discovered and bound at runtime. The context model in [5] itself however is directly consumed by the context aware system, e.g. for deciding about a certain reconfiguration based on a specific context state. Thus the adaptation logic and certain context dependencies are hard coded into the adaptive system. Hence the logic and results can not be shared among several subsystems and moreover the adaptation can not be reconfigured itself (e.g. to meet varying CPU resources).

The model proposed here, and illustrated in Figure 2, therefore extends the context model as defined in [5] by adding not only sources and computational nodes, but also sinks for contextual information. Adaptation like any other usage of context information becomes visible (and also detectable and replaceable at runtime) as actuator elements. *Actuators* represent parts of the system that access or observe certain parts of the context [10]. Additionally, it is possible to move adaptation logic into such a context adaptation model, instead of being hidden inside the actuators, if the result of an adaptation decision is just modeled as another context, describing how the system should look after the necessary reconfigurations (reconfiguration context). With this extension, adaptation can be defined as special interpreters that take information from the initial or intermediate context data elements and compute a specification of how the new system should look after the adaptation. The complete model for adaptation (context awareness) in this document thus is made up of all context information, the sources for information in form of sensors, any decision or interpretation in form of interpreters and at least one actuator that reconfigures the system according to the description found in the reconfiguration context data. In the following, a recapitulation of the four basic elements of the model is provided.

Sensors and Context Elements. Sensors observe the system's environment and collect data from the real world (environment). This data is gathered as context elements in a context server. A *velocity sensor* of a car for instance measures the context element *absolute velocity*.

Interpreters. Interpreters observe contexts and update contexts stored in profiles. They retrieve context elements, analyze the including information and/or compute new information. Thus, they produce decision context elements and the suggested target state of the system. E.g. the velocity interpreter analyzes the situation, perhaps based on further context elements, and computes a new configuration such as *displaying the velocity with a larger font*.

Actuators. Actuators observe the adaptation context, and if required adapt the system according to the identified situation. They activate the new configuration based on interpreters' propositions (actuators can manage reconfigurations, provide information [18] or trigger command messages [22] to further technical components).

2.4 The Overall Process of Adaptation

Summarizing section 2.3, the overall process of adaptation can be viewed as a three step process, namely monitoring the environment of a system (detecting context), followed by choosing the best appropriate configuration (deciding about new configurations) and finally deploying a new configuration at runtime (i.e. reconfiguring). The deployment step can involve for example changing behavior, implementations, modifying structure, adding or removing functionality or even downloading new code. Hence adaptation can be modeled as a simple subsystem consisting of the four base elements (sensors, interpreters, actuators, context). Adaptive Systems therefore logically contain three essential parts: Adaptation Logic, Core System (Minimal necessary system functionality) and System Environment (containing fluctuating components used to extend system functionality). This and regarding adaptation as a separate subsystem eases formal modeling of adaptation functionality as a filter. A formalization of adaptation based on the FOCUS formalism [2] was presented in [14] and [7]. There are however no formal system boundaries separating environment, core and adaptation. This separation is always an engineering decision based on specific requirements regarding necessary flexibility, security, safety etc.).

In the CAWAR framework [14], [15], Sensors, interpreters and actuators are realized as abstract classes. Before their usage the abstract classes must be implemented depending on the concrete interpretation (e.g. a rule, a neural network etc.) or actuation (a certain device, telecommunication service/layer, application etc.) functionality. To allow for adaptable adaptation again components derived from these classes are used in form of a reconfigurable service represented by a transparent service proxy component. This proxy component is part of the context adaptation subsystem (not part of the adaptable/reconfigurable core subsystem). The functionality itself however, i.e. the component bound to the service proxy at runtime can be either part of the system core or the technical system environment.

3   The Approach of Calibration

Section 2 described that adaptation is technically based on reconfigurability, for example changing behavior, changing realization (implementation), modifying structure, adding/removing functionality or downloading new parameters or even code.

Realizing adaptation of a system using such reconfigurability controlled by context information handling and decision logic that were designed at development time works quite well for small and specialized adaptive applications within relatively stable and predictable environments like for example a prototype in a test lab. There are even indications that there is a group of context aware application scenarios that are especially tolerant toward wrong adaptation by nature, because their results can be easily ignored, if wrong, while still deemed helpful by the user, if right. An example here would be a news or a weather terminal. Most published research works in the domain of ubiquitous computing fall into either of these two categories (i.e. [6], [17], [20] or [21]).

Own experiments [7] with adaptive systems not belonging to one of these two categories yet, confirmed that complex real world adaptive systems are usually supposed to very likely fall into the trap of the frame problem (see [12] and [4]).

This is a well known problem from AI about the difficulties describing an infinite complex and dynamically changing world using static assumptions (i.e. models). Over time some of these assumptions and therefore abstractions used in a model can get wrong, even if they were valid while constructing the model. This again leads to false (compared to reality) decisions [19] and therefore will result in spontaneous unwanted behavior (SUB). SUB is not an error in the traditional way [1] since there is no reasonable preventable mistake that led to the SUB, for example the emerging of new user needs, changed laws, new device types etc. This problem is very common in ubiquitous computing because the user moves through a wide spread landscape of situations with very different requirements and these requirements can not all be gathered at design time or are very likely to change over time, by location or preferences of the user. A good example is an intelligent fridge that can not understand that the presently contained food is only for tomorrows special party and does not need to be reordered if used up. Even if this problem is not solvable in general, it can be avoided or at least circumvented by changing the model that reasons about reality (i.e. our context adaptation) from time to time to meet the changed reality. An other way is to have the user compensate for the changed reality by adapting his behavior, knowledge or expectations so that the adaptive system still can produce correct results, even if the real world has changed outside the scope of initial requirements and parameters that were anticipated by the system's developers.

Either-way, these corrective processes are called *calibration* and can be seen as an adaptation of the adaptation itself. However it is obvious that with state of the art adaptation and context awareness concepts, there is always part of the model that can not be changed like the logical service architecture, decision logic, context management etc. Usually this not changeable part of an adaptive system is the adaptation deciding and controlling subsystem in contrast to the adaptable core subsystem, which is reconfigurable by the way adaptation was designed, i.e. exposing certain switchable behavior subsets to all observers beyond the system boundaries of the adaptive system (*partial reconfigurability*).

3.1 Adapting the Adaptation Logic

To avoid SUB effects from static world models our approach of adaptation is mainly focused on reconfiguration of the adaptation subsystem. This means that the adaptation logic itself can be target of context dependent adaptation decisions. Combined with the partial reconfigurability achieved by adaptation of the reconfigurable system core we therefore get an implicitly total reconfigurable system. To achieve such a *total reconfigurability*, we regard a description of the adaptation with its world model and decision logic itself as contextual information that can be modified by sensors and interpreters and then used by a special actuator called *activator* to reconfigure the adaptation and context of a system (e.g. by adding a new context or decision logic at runtime). To better understand the process of calibration, it is helpful to recall the role of adaptation context (see section 2.3). All decisions and transformations will result in an *adaptation context*. This is the description of how the system should look after successful adaptation and this is accordingly what the activator will change the adaptation subsystem if this adaptation context contains a reflectable model of the system's adaptation.

Such a model of context adaptation that can describe its self reconfiguration is called calibrateable model (*k-model*). It is obvious with this definition that any specific implementation of a k-model with an actuator that can read and reconfigure context adaptation models can serve as a generic framework, because it can be fed with any other specific specification of a context adaptation and still will reconfigure itself accordingly due to its total reconfiguration ability. However this means a common formal founded semantic of the k-model is necessary to ensure that every refinement of our abstract k-model into a technical implementation is compatible with others it can be reconfigured into. For this purpose a mathematical founded base model which consists of components and channels describing mathematical functions processing sets of infinite message streams [2] is used. In this base model, systems or subsystems are described as a network of components that communicate with each other over channels. Their behavior is specified as a relation between communication histories of input and output channels. A communication history is expressed as a stream of messages. The relation between input and output message histories describes a specification of visible component behavior.

Adaptation in this basic formal model is interpreted as a change of network components representing the adapted system, i.e. components or channels can be added or removed resulting in a change of visible behavior of the system. In principle any model results from abstraction using static assumptions. Since models hence are static approximations of reality, dynamics can at best be emulated. Adaptation therefore can only be formalized using a superposition of all

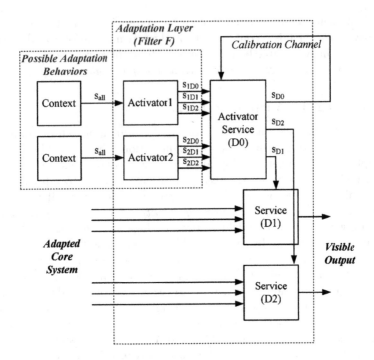

Fig. 3. Calibrateable Adaptation Schematics

possible structures and functionality as well as a behavior to emulate a system reconfiguration by switching back and forth between the configuration possibilities. One could argue that this way it is not possible to formalize adaptation because it is not possible to reconfigure a system into a state that is unknown at specification time of the system, e.g. by downloading new software modules at runtime.

However there is a slight but important difference between the specification of a system and its implementation. While the implementation is part of reality, a specification is a model of this and maybe similar realities by restricting a state of all possible systems down to a usually still infinite group of systems showing the expected behavior (the static approximation mentioned above). Adaptation can thus easily be modeled as a superposition of states without needing to enumerate or even know each and every possible adaptation state as long as the switching behavior between them can be expressed. In the case of the formal base model, this can be defined as schematics mathematically describing the relation between communication histories of a set of typed input and output channels that can filter the output of certain components or channels that are not active in a certain adaptation state (see Fig. 3, the D0 component filtering out one of two alternative adaptation configurations that control the reconfiguration of the core). Looking at the formalization it becomes clear that apparently context adaptation is a purely engineering construct; since no extra functionality is added (we still only have channels and components). Structuring certain behavior changes in a system the way it is expressed within the abstract context adaptation model and its formal foundation, however has some clear advantages from a systems engineering point of view dealing with system flexibility:

- A clear segmentation and precise switching allowing for running a system that is only partially implemented. Precise adaptation makes sure no specified behavior is exposed that is not implemented yet.
- Decoupling communication between certain components such as sensors using the context allows for communication of components that not even exist at the same time or which have an availability (like in wireless or mobile networks) that can not be controlled by the system itself. So even if a sensor for needed information is not available at the moment, its last known value still might be available from the context.
- Context awareness defines a basic interface for sensor-, interpreter-, and actuator components ensuring that all current and future components can establish a very basic communication with each other (i.e. exchanging context data). Based on this, enhanced protocols can be negotiated. This allows for easy expansion of a system with a priori unknown functionalities.

3.2   Calibrateable Adaptation Schematics

Seeing adaptation as an engineering construct only makes sense in conjunction with a technical reconfiguration possibility to automatically implement a specified system. In software engineering there are several concepts for dynamically implementing a software system (e.g. late binding, DLLs etc.). The most flexible concept to date however, are services that even allow for changing an active implementation at runtime. The service itself is specified as a group of similar behavior while abstracting component (implementation) identities. A network of services therefore is sometimes referred to as a logical architecture defining static interdependencies between all possible implementations that could be exchanged at runtime. This allows automatic implementation of a system from a set of currently available implementations at runtime. Since more than one component implementation can be used to implement a given service, this process is sometimes referred to as Design@Runtime [9].

Services are an sufficient technical concept to implement our mathematical model of adaptation since services usually are realized using a proxy access component that can act as a switch between several component implementations and therefore acts like the adaptation filter component (actuator) of our abstract model. However changing the system by switching component implementations has some invariants in form of the logical architecture (services), i.e. the fulfilled function, task or requirements of the given system or subsystem. Restricting adaptation to the use of services therefore only produces a partial reconfigurable system, which is prone to be affected by the aforementioned frame problem.

To technically realize total reconfiguration of a system using adaptation the service concept therefore needs to be extended to also change the logical architecture using the activator technically mentioned before. The activator is itself a service that controls the reconfiguration channels of all possible service proxies in a given system. The activator can set the component implementation that is used by a service proxy or can switch off the service proxy by deactivating its output channels that are observable from the outside of the system. The activator also controls its own service proxy. This way it can hand over the reconfiguration control to any other activator component implementation achieving a total reconfiguration without any invariants if necessary.

Using the activator, any system's adaptation behavior can not only be calibrated but also be bootstrapped (i.e. calibrated from an empty adaptation). A description of the application's adaptation is loaded from an outside source into the context server using the adaptation model sensor (e.g. loading from a file). A special actuator (the model actuator) reads the adaptation model description from the context. Since this model actuator is implemented as an activator it can activate and bind any further sensor, interpreter, actuator and context service described in the adaptation model description loading at bootstrap time.

At any later time it is possible to modify (calibrate) this adaptation description stored in its own context as long as the components defining the bootstrap adaptation are still present and active or have been replaced by implementations of the same functionality. The model of a concrete adaptation behavior stored in the context is described as an XML document. It contains information about all sensors, interpreters, context elements and actuators described as services with their ids, syntactic and semantic type information. The syntactic type is usually composed from an interface description (IDL, WSDL etc.) but can also contain binding information like visible behavior or even the reference of a specific component instance. The semantic type, in contrast, is used to distinguish technical identical system components, like for example an inside and an outside temperature sensor, or can be used for ontology based searches without specifying a concrete interface. This can work out well for this special case, because as already mentioned the abstract adaptation model only consists of four basic roles that have a minimum common communication API (i.e. accessing context information). So it is possible for example to search and bind two components with unknown interfaces. It just needs to be made sure that they can communicate with each other based on context datagram. A short summary of the technical realization of the framework, the mathematical model and its formal description method was presented in [14].

4   Graphical Notation for Adaptation Models

The XML document used to describe such flexible adaptation models can moreover be easily mapped onto the formally founded mathematical model. This allows for a wide range of runtime checks of such specifications, for example testing for consistency before deploying new adaptation behavior. There are also syntactical mappings between the text based XML description and a graphical notation, described in [7]. Its main purpose besides being useful when designing tool support, is being the basis for documenting adaptation behavior toward

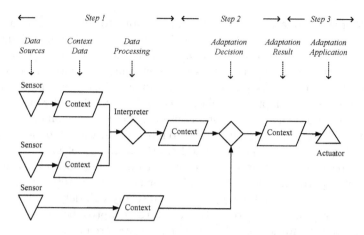

Fig. 4. A Graphical Notation for Adaptation Models

users that can then use this knowledge for calibrating the adaptation toward their personal preferences. To support this purpose the given framework furthermore contains a set of specialized syntactic and optical transformers that can modify the original graphical specification for better clearness and design, for example by automatically generating indexed sequencings of complex models or information folding/unfolding techniques. Figure 4 illustrates a graphical notation for the model, depicting the three steps of adaptation.

5 Conclusion and Future Works

So far, context models for adaptability were incomplete and neither flexible nor modular enough. Incomplete because they mainly emphasize deriving context data from a system situation and disregard the opposite direction of adapting the system situation based on the context data.

Besides sensors and interpreters like in [5], we therefore introduced the concept of an adaptation context as a characterization of how the system should look like after an adaptation activity carried out by actuators. Existing models are not flexible enough, because context is defined regarding a system situation but the exact relation between situations and context remains unclear [12]. Therefore, context in our case is an operationalization that is used to detect a certain situation. This kind of model of a situation is usually developed by the context awareness designer and is of static nature. A later customization or modification of this kind of context aware behavior to meet new conditions that were not foreseen by the system's designer is difficult. The very complex world it describes, however, continues to change over the lifetime of the system. This problem, also known as the frame problem in AI research [4], would make any context aware system and hence any adaptations fail sooner or later.

We sketched out that it is possible to avoid this problem if it would depend on the situation what is an adequate context model. Our concept of an adaptation context therefore can also, besides adaptations done to the core, hold a description of changes in the specification of context dependent adaptation behavior of the system itself. The process of changing context or adaptive behavior depending on the current context is not necessarily self-contained since the information can be produced by sensors and interpreters outside of the system. This way the system's context and adaptation behavior can also be modified by means outside the scope and knowledge of the system designer during development time of the system, thus providing enough flexibility to integrate further modular techniques (including human customization) to address the frame problem.

Further research effort needs to be put into applying and integrating various concrete approaches from AI research for changing adaptive behavior and relevance. What is also still missing is a semantically well defined process to design adaptive systems, while concentrating on the adaptive behavior rather than discussing implementation details. Our ongoing research suggests an approach that involves facing possible abstract situations with objectives and afterwards designing an intermediary connection between the two. This is done in form of formal conditions and action specifications to reach the planned situation. Both conditions and actions are then expressed as context interpreters inside the context model itself.

References

1. Breitling, M.: Formale Fehlermodellierung für Verteilte Reaktive Systeme. Dissertation, Technische Universität München, Fakultät für Informatik, 2001.
2. Broy, M., Stølen, K.: Specification and Development of Interactive Systems - Focus on Streams, Interfaces and Refinement. Monographs in Computer Science, Springer-Verlag, 2000.
3. Chan, A., Chuang, S.: MobiPADS: A Reflective Middleware for Context-Aware Mobile Computing. IEEE Transactions on Software Engineering, 29(12), 2003.
4. Dennett, D.: Cognitive Wheels: The Frame Problem of AI. In C. Hookway, editor, Minds, machines, and evolution, pages 129-151. Cambridge University Press, Cambridge, 1984.
5. Dey, A. K.: Providing Architectural Support for Building Context-Aware Applications. PhD thesis, College of Computing, Georgia Institute of Technology, 2000.
6. Dey, A. K., Abowd, G. D.: CybreMinder: A Context-Aware System for Supporting Reminders. In Proceedings of the 2nd International Symposium on Handheld and Ubiquitous Computing (HUC2K), pp. 172-186, Bristol, UK, September 25-27, 2000.
7. Fahrmair, M.: Kalibrierbare Kontextadaption für Ubiquitous Computing. Dissertation, Technische Universität München, 2005.
8. Fahrmair, M., Mohyeldin, E., Salzmann, C.: Communication Profiles for Reconfigurable Systems. In: Dillinger, Madani, Alonistioti (Eds.), Software Defined Radio: Architecture, Systems and Functions. John Wiley & Sons, 2003
9. Fahrmair, M., Salzmann, C., and Schoenmakers, M.: A Reflection Based Tool for Observing JINI Services. In Reflection and Software Engineering, number 1826 in LNCS. Springer-Verlag, 2000.

10. Houssos, N., Alonistioti, A., Merkakos, L., Mohyeldin, E., Dillinger. M., Fahrmair, M., Schoenmakers, M.: Advanced Adaptability and Profile Management Framework for the Support of Flexible Mobile Service Provision. IEEE Wireless Communications Mag, 2003.

11. Lieberman, H., Selker, T.: Out of Context: Computer Systems that Adapt to, and Learn from, Context. IBM Systems Journal, 39(3-4): 617-632, 2000.

12. Lueg, C.: Operationalizing Context in Context-Aware Artifacts: Benefits and Pitfalls. Informing Science, 5(2), 2002.

13. Merriam-Webster: Collegiate Dictionary. Merriam-Webster, Inc., 2003.

14. Mohyeldin, E., Dillinger, M., Fahrmair, M., Sitou, W., Dornbusch, P.: A Generic Framework for Negotiations and Trading in Context Aware Radio. In Software Defined Radio Technical Conference, Phoenix Arizona USA, 2004.

15. Mohyeldin, E., Fahrmair, M., Sitou, W., Spanfelner, B.: A Generic Framework for Context Aware and Adaptation Behaviour of Reconfigurable Systems. In 16th IEEE International Symposium on Personal Indoor and Mobile Radio Communications, Berlin, Germany, 2005.

16. Mohyeldin, E., Schulz, E., Dillinger, M., Fahrmair, M., Dornbusch, P. Dynamic Reconfiguration of Wireless Middleware. In: IST Mobile & Wireless Communications Summit 2004, Lyon/France, IST, 2004

17. Mozer, M. C.: Lessons from an Adaptive House. In Cook, D. Das, R. (Eds.), Smart Environments: Technologies, Protocols, and Applications. J. Wiley & Sons. 2004.

18. Pascoe, J.: Adding Generic Contextual Capabilities to Wearable Computers. In Proceedings of the 2nd IEEE International Symposium on Wearable Computers (ISWC98), pp. 92-99, Pittsburgh, PA, IEEE, October 19-20 1998.

19. Pfeifer, R., Rademakers, P.: Situated Adaptive Design: Toward a Methodology for Knowledge Systems Development. In W. Brauer, D. Hernandez, editors, Proceedings of the Conference on Distributed Artificial Intelligence and Cooperative Work, pages 53-64. Springer Verlag, 1991.

20. Rhodes, B.: The Wearable Remembrance Agent: A System for Augmented Memory. In Personal Technologies Journal Special Issue on Wearable Computing, 1:218-224, Personal Technologies, 1997.

21. Sawhney, N., Wheeler, S., Schmandt, C.: Aware Community Portals: Shared Information Appliances for Transitional Spaces. In Journal of Personal and Ubiquitous Computing, Vol. 5 Issue 1, pp. 66-70. Springer-Verlag, February 2001.

22. Schilit, W. N.: System Architecture for Context-Aware Mobile Computing. PhD thesis, Columbia University, 1995.

23. Sutcliffe, A., Fickas, S., Sohlberg, M.: Personal and Contextual Requirements Engineering. 13th IEEE International Conference on Requirements Engineering, Paris, September 2005.

Author Index

Lecture Notes in Artificial Intelligence (LNAI)

Vol. 3690: M. Pěchouček, P. Petta, L.Z. Varga (Eds.), Multi-Agent Systems and Applications IV. XVII, 667 pages. 2005.

Vol. 3684: R. Khosla, R.J. Howlett, L.C. Jain (Eds.), Knowledge-Based Intelligent Information and Engineering Systems, Part IV. LXXIX, 933 pages. 2005.

Vol. 3683: R. Khosla, R.J. Howlett, L.C. Jain (Eds.), Knowledge-Based Intelligent Information and Engineering Systems, Part III. LXXX, 1397 pages. 2005.

Vol. 3682: R. Khosla, R.J. Howlett, L.C. Jain (Eds.), Knowledge-Based Intelligent Information and Engineering Systems, Part II. LXXIX, 1371 pages. 2005.

Vol. 3681: R. Khosla, R.J. Howlett, L.C. Jain (Eds.), Knowledge-Based Intelligent Information and Engineering Systems, Part I. LXXX, 1319 pages. 2005.

Vol. 3673: S. Bandini, S. Manzoni (Eds.), AI*IA 2005: Advances in Artificial Intelligence. XIV, 614 pages. 2005.

Vol. 3662: C. Baral, G. Greco, N. Leone, G. Terracina (Eds.), Logic Programming and Nonmonotonic Reasoning. XIII, 454 pages. 2005.

Vol. 3661: T. Panayiotopoulos, J. Gratch, R.S. Aylett, D. Ballin, P. Olivier, T. Rist (Eds.), Intelligent Virtual Agents. XIII, 506 pages. 2005.

Vol. 3658: V. Matoušek, P. Mautner, T. Pavelka (Eds.), Text, Speech and Dialogue. XV, 460 pages. 2005.

Vol. 3651: R. Dale, K.-F. Wong, J. Su, O.Y. Kwong (Eds.), Natural Language Processing – IJCNLP 2005. XXI, 1031 pages. 2005.

Vol. 3642: D. Ślęzak, J. Yao, J.F. Peters, W. Ziarko, X. Hu (Eds.), Rough Sets, Fuzzy Sets, Data Mining, and Granular Computing, Part II. XXIII, 738 pages. 2005.

Vol. 3641: D. Ślęzak, G. Wang, M. Szczuka, I. Düntsch, Y. Yao (Eds.), Rough Sets, Fuzzy Sets, Data Mining, and Granular Computing, Part I. XXIV, 742 pages. 2005.

Vol. 3635: J.R. Winkler, M. Niranjan, N.D. Lawrence (Eds.), Deterministic and Statistical Methods in Machine Learning. VIII, 341 pages. 2005.

Vol. 3632: R. Nieuwenhuis (Ed.), Automated Deduction – CADE-20. XIII, 459 pages. 2005.

Vol. 3630: M.S. Capcarrère, A.A. Freitas, P.J. Bentley, C.G. Johnson, J. Timmis (Eds.), Advances in Artificial Life. XIX, 949 pages. 2005.

Vol. 3626: B. Ganter, G. Stumme, R. Wille (Eds.), Formal Concept Analysis. X, 349 pages. 2005.

Vol. 3625: S. Kramer, B. Pfahringer (Eds.), Inductive Logic Programming. XIII, 427 pages. 2005.

Vol. 3620: H. Muñoz-Ávila, F. Ricci (Eds.), Case-Based Reasoning Research and Development. XV, 654 pages. 2005.

Vol. 3614: L. Wang, Y. Jin (Eds.), Fuzzy Systems and Knowledge Discovery, Part II. XLI, 1314 pages. 2005.

Vol. 3613: L. Wang, Y. Jin (Eds.), Fuzzy Systems and Knowledge Discovery, Part I. XLI, 1334 pages. 2005.

Vol. 3607: J.-D. Zucker, L. Saitta (Eds.), Abstraction, Reformulation and Approximation. XII, 376 pages. 2005.

Vol. 3601: G. Moro, S. Bergamaschi, K. Aberer (Eds.), Agents and Peer-to-Peer Computing. XII, 245 pages. 2005.

Vol. 3600: F. Wiedijk (Ed.), The Seventeen Provers of the World. XVI, 159 pages. 2006.

Vol. 3596: F. Dau, M.-L. Mugnier, G. Stumme (Eds.), Conceptual Structures: Common Semantics for Sharing Knowledge. XI, 467 pages. 2005.

Vol. 3593: V. Mařík, R. W. Brennan, M. Pěchouček (Eds.), Holonic and Multi-Agent Systems for Manufacturing. XI, 269 pages. 2005.

Vol. 3587: P. Perner, A. Imiya (Eds.), Machine Learning and Data Mining in Pattern Recognition. XVII, 695 pages. 2005.

Vol. 3584: X. Li, S. Wang, Z.Y. Dong (Eds.), Advanced Data Mining and Applications. XIX, 835 pages. 2005.

Vol. 3581: S. Miksch, J. Hunter, E.T. Keravnou (Eds.), Artificial Intelligence in Medicine. XVII, 547 pages. 2005.

Vol. 3577: R. Falcone, S. Barber, J. Sabater-Mir, M.P. Singh (Eds.), Trusting Agents for Trusting Electronic Societies. VIII, 235 pages. 2005.

Vol. 3575: S. Wermter, G. Palm, M. Elshaw (Eds.), Biomimetic Neural Learning for Intelligent Robots. IX, 383 pages. 2005.

Vol. 3571: L. Godo (Ed.), Symbolic and Quantitative Approaches to Reasoning with Uncertainty. XVI, 1028 pages. 2005.

Vol. 3559: P. Auer, R. Meir (Eds.), Learning Theory. XI, 692 pages. 2005.

Vol. 3558: V. Torra, Y. Narukawa, S. Miyamoto (Eds.), Modeling Decisions for Artificial Intelligence. XII, 470 pages. 2005.

Vol. 3554: A.K. Dey, B. Kokinov, D.B. Leake, R. Turner (Eds.), Modeling and Using Context. XIV, 572 pages. 2005.

Vol. 3550: T. Eymann, F. Klügl, W. Lamersdorf, M. Klusch, M.N. Huhns (Eds.), Multiagent System Technologies. XI, 246 pages. 2005.

Vol. 3539: K. Morik, J.-F. Boulicaut, A. Siebes (Eds.), Local Pattern Detection. XI, 233 pages. 2005.

Vol. 3538: L. Ardissono, P. Brna, A. Mitrović (Eds.), User Modeling 2005. XVI, 533 pages. 2005.

Vol. 3533: M. Ali, F. Esposito (Eds.), Innovations in Applied Artificial Intelligence. XX, 858 pages. 2005.

Vol. 3528: P.S. Szczepaniak, J. Kacprzyk, A. Niewiadomski (Eds.), Advances in Web Intelligence. XVII, 513 pages. 2005.

Vol. 3518: T.-B. Ho, D. Cheung, H. Liu (Eds.), Advances in Knowledge Discovery and Data Mining. XXI, 864 pages. 2005.

Vol. 3508: P. Bresciani, P. Giorgini, B. Henderson-Sellers, G. Low, M. Winikoff (Eds.), Agent-Oriented Information Systems II. X, 227 pages. 2005.

Vol. 3505: V. Gorodetsky, J. Liu, V.A. Skormin (Eds.), Autonomous Intelligent Systems: Agents and Data Mining. XIII, 303 pages. 2005.

Vol. 3501: B. Kégl, G. Lapalme (Eds.), Advances in Artificial Intelligence. XV, 458 pages. 2005.

Vol. 3492: P. Blache, E.P. Stabler, J.V. Busquets, R. Moot (Eds.), Logical Aspects of Computational Linguistics. X, 363 pages. 2005.